Electronic
BRAINS

Electronic BRAINS

STORIES FROM THE DAWN OF THE COMPUTER AGE

MIKE HALLY

Joseph Henry Press
Washington, D.C.

Joseph Henry Press • 500 Fifth Street, NW • Washington, DC 20001

The Joseph Henry Press, an imprint of the National Academy Press, was created with the goal of making books on science, technology, and health more widely available to professionals and the public. Joseph Henry was one of the founders of the National Academy of Sciences and a leader in early American science.

Any opinions, findings, conclusions, or recommendations expressed in this volume are those of the author and do not necessarily reflect the views of the National Academy of Sciences or its affiliated institutions.

Library of Congress Cataloging-in-Publication Data

Hally, Mike.
  Electronic brains : stories from the dawn of the computer age / Mike Hally.
    p. cm.
  Includes bibliographical references and index.
  ISBN 0-309-09630-8 (cloth)
  1. Computers—History. I. Title.
  QA76.17.H35 2005
  004--dc22

                          2005016583

Cover design by Michele de la Menardiere; photo, copyright by Hulton-Deutsch Collection/CORBIS

Printed in the United States of America.

CONTENTS

v

PREFACE

This book is about that brief period just after World War II when the first truly modern computers were developed by a range of pioneering teams across four continents. The book grew out of a series of four 15-minute programs called *Electronic Brains* on BBC Radio 4 that received an unexpectedly warm welcome from listeners who had had no idea that computer history could interest them. It did that by avoiding too much technical detail, instead exploring what it was actually like to be one of the pioneers in the early computer projects. To get those accounts meant finding surviving members of the pioneering teams in various parts of Britain, America, Australia, and the former Soviet Union, and recording their memories at some length, memories that were invariably clear and vivid.

You can't fit much of that material into an hour's radio—about as much as a single chapter in this book—so this version of *Electronic Brains* is much more than just transcripts of the radio programs. There is more about each of the four stories from the original series, the contributors have more space to speak for themselves and there is more historical context. There is a chapter about the early Australian computers, unjustly overlooked in most accounts. And a major question that arises out of those stories—why IBM became so dominant despite a late entry into the field—is explored in its own chapter.

My first experience of computers moving out of the laboratory and into everyday life came in the early 1970s in a large aerospace engineering company. A shiny new hot-drinks machine replaced the old one on the shop floor and it had a numbered keypad instead of large blue plastic buttons. It had a simple computer inside it, so instead of pushing the button marked "coffee white no sugar" we had to press the numbers "1, 4, 3" on the keypad. It baffled us and for several days we kept getting the wrong drinks, though we were all electronic engineers and we knew that microcomputers were going to find their way into all kinds of products. Within a few years many of us had some sort of computer at home, and they were fascinating to us even though there really wasn't that much you could do with a Sinclair Z80, other than marvel at it.

Recalling that period helps me understand why all the people I interviewed for this book so obviously enjoyed their work, which was intellectually demanding with long hours and scarce resources. As one of the LEO computer engineers, Frank Land, remarked, "It was very exciting because everything you did had never been done before." The work was the technological equivalent of climbing Everest or walking to the South Pole, and there was a huge sense of camaraderie, of sheer fun, in those early days, which comes across strongly in the memories of these men and women.

Bringing these accounts together allows certain themes to emerge. One of the most striking is the effect of World War II on the development of the early computers. In some cases the onset of war held back progress, but in most it accelerated it. The war didn't just bring into being major military projects like the British code-breaking Colossus and the American ballistic computer the ENIAC; there were indirect effects as well. The British professor Sir Maurice Wilkes said the biggest lesson he learned from wartime radar work was "how to get things done." Across the Atlantic the legendary boss of IBM, Thomas J. Watson Sr., saw that the urgency of wartime production would force a similar increase in the pace of technological change in his country when peace returned. Another important effect of the war in the United States was the large number of young men who got technical training through the "GI Bill," which entitled ex-servicemen to a free college education. Many of the engineers working on the early American computers could not have done so without the qualifications they acquired through the GI Bill.

What is also noteworthy is the way the first computers emerged more or less simultaneously in various parts of the globe, in a general climate of technological advancement that spread across the developed world. The great Soviet pioneer Sergei Lebedev may have been spurred on by news reports of the American ENIAC in 1946, but he had been experimenting with the principles of digital computing some years earlier and the machine his team created was all their own design. So there is no simple answer to the question, "Who invented the computer?" It was a process of development across several continents that involved thousands of people. Certainly some individuals made outstanding contributions and others were exceptionally far-sighted, but no one has the undisputed title of "Father of the Computer." John Atanasoff, with a better claim than most as creator of the Atanasoff Berry Computer (ABC), summed it up well when he said, "There is enough credit for everyone in the invention and development of the electronic computer."

Atanasoff was quite typical in being a professor of both mathematics and physics. While some of the pioneers were particularly renowned as brilliant mathematicians, like Alan Turing, or physicists, like John Mauchly, nearly all had expertise in both disciplines. It is interesting that most of the early projects were led by teams of two. Atanasoff needed the gifted electronics student Clifford Berry to make his design a reality, just as Mauchly needed

Presper Eckert. In Australia the mathematician and physicist Trevor Pearcey found his electronics guru in Maston Beard. The Russian Sergei Lebedev was almost unique in having the required level of expertise in mathematics, physics, and electronics. Women played significant roles in nearly all the early projects, whatever the country and political system. Female mathematicians, whose employment opportunities had generally been more limited than those of their male counterparts, found their skills in demand and often showed particular aptitude for the new art of programming.

Because of the global coverage and the extensive use of direct quotes in this book I've adopted local usage where appropriate. So in the American and Australian chapters I write of electronic "tubes," while in the British accounts the same devices are called "valves." Similarly, in the account of the first Soviet computer, I've preserved the use of English of the Ukrainian speakers rather than rewriting their testimony in standard English, and I've used their preferred spelling of Kyiv rather than the older Kiev. Many of the quotes in this book are taken from recent interviews with pioneers looking back half a century or more, and I indicate this by using the present tense. Extracts from contemporaneous accounts are introduced in the past tense.

It only remains to thank all those who have assisted so generously in making this book possible: Gail Lynch at Granta, who heard a quirky little radio series and thought it might be the basis for a book; Mark Whitaker, who presented that series, conducted the American interviews and cast his historian's eye over the scripts; all the surviving pioneers of these early projects who granted me interviews and conveyed their experiences so vividly; David Caminer and Frank Land, who read the LEO chapter and corrected some glaring errors; Erik Rambusch, Andrew Egendorf, and Bill Wenning, who did the same for the Rand 409; Viktor

Ivaneko, who led me to the pioneers in the Ukraine—I'd have been lost without him; John Deane, who drew my attention to the early Australian computers and generously gave me his contacts; and above all Doron Swade, who read and forensically scrutinized the whole manuscript despite many other competing demands on his time. Naturally any remaining errors are my responsibility.

Mike Hally
author@electronicbrains.info

PROLOGUE

Counting is surely one of the earliest human intellectual activities and the first counting aids must have been fingers, closely followed by piles of stones, and then tally-marks on cave walls and the like. The wonderfully simple abacus evolved from its earliest-known form in Babylonia in the last millennium BC, and enjoyed a useful life well into the second millennium AD. It is a true design classic and anyone who grew up in the 1950s probably had an abacus built into one side of their playpen. More elaborate counting machines began to appear in the centuries to follow, such as Leonardo da Vinci's fifteenth-century design in his manuscript the "Codex Madrid" and Blaise Pascal's seventeenth-century "Arithmetic Machine." There was a fundamental problem for these devices, which may be of some consolation to those who struggled with basic arithmetic as children, and that was the need to "carry" numbers from the units to the tens column and so on. Even a mechanism that would automatically perform a single carry (as in the addition of 6 and 7 to get 13) would struggle with a whole string of carries (say adding 5 and 999997 to get 1000002).

The Englishman Charles Babbage is widely credited with being the first person to employ principles that would much later be recognized as fundamental to electronic computing. He was one of many people who were frustrated by the numerous errors in the mathematical tables that had become widely available in the eighteenth century. This wasn't just an academic problem. Tide tables and star charts in particular were calculated by hand, perhaps with the aid of slide-rules (invented a couple of centuries earlier), but such tables were notoriously unreliable, with mistakes in calculation, transcription, and printing. These were serious matters in an age when international trade depended largely on maritime transport and errors cost ships, goods, and lives. It is said that Babbage, a great collector of such tables, one day cried out in exasperation that he wished they had been calculated by steam power and then set about designing a machine to do the job.

Only mathematicians could solve the equations that yielded functions like logarithms and sines, but a simplified procedure known as the "method of differences" gave results that were accurate enough when carried out properly. This method could be used by teams of people with only basic arithmetic skills and Babbage realized it could also be mechanized to eliminate the human errors. So he called his calculating machine the "Difference Engine" and started work on it in 1821. Results would be embossed directly on to thick sheets that could be used to print the final tables, thus eliminating the other sources of human error in production. Unfortunately the whole design called for the manufacture of some 25,000 parts, most of them at the limit of materials and machining tolerances of the time. After 11 years, the expenditure of a great deal of government money, and a terminal dispute with his engineer, work stopped. Eventually a small part of the Difference Engine was completed; it worked well (and functions impeccably to this day, according to the Science Museum in London, where it is now kept) and can be regarded as the first automatic calculator.

Although it proved a big hit at dinner parties (Babbage was a great socialite), the incomplete engine didn't produce the accurate printed tables Babbage wanted. That was achieved by a father-and-son team, Georg and Edvard Scheutz, who read about Babbage's efforts and then designed and built their own machine in Sweden. Scheutz senior forecasted in 1833 that just one such machine "would suffice for the needs of the whole world," probably the first example of a prominent figure conspicuously underestimating the potential of computing. However, the Scheutzes, who built their prototype in just six years with only modest machining facilities, had trouble persuading the world that it needed any such machine at all. This was a pity because, although the prototype was rather basic, it worked and it printed the results. They produced a few more machines to a higher standard of accuracy, which also worked but were rather unreliable and didn't really produce the expected benefits. Both father and son died bankrupt, six years apart.

Babbage meanwhile had started pursuing another project, his "Analytical Engine," and this really earned him a place in computer history. During his years of work on the Difference Engine, he had conceived of a far more powerful machine that would be universal in the sense that it would be capable of solving a variety of algebraic equations. Just as he had originally called for mathematical tables to be calculated by steam, he looked again to the technologies of the day for inspiration and to the textile industry in particular. He designed machinery to add, subtract, multiply, and divide, and he called this the "mill"; this was analogous to the arithmetic processor in an electronic computer. Rather than duplicate this machinery all over the engine, he intended the mill to be a single central mechanism that would fetch data from another part of the engine he called the "store." He came up with a method for the engine to carry out various kinds of calculations by following a set of instructions that defined the equations to be solved. This was long before magnetic tape was invented, but there was a suitable technology, developed for the textile industry: punched cards. These had recently been invented by Joseph-Marie Jacquard, a Frenchman who used holes punched in cards to control the machines that wove intricate patterns in silk (the idea was also adapted for the mechanical piano, the pianola). So Babbage used punched cards to define the instructions—the "program," as we would call it—for his Analytical Engine. Thus the Analytical Engine would carry out a series of discrete steps, just as a digital computer does today.

Babbage also gave his engine the ability to choose one set of instructions over another depending on the result of an earlier calculation, what we now call "conditional branching." This was a remarkable feature, not present in many of the early electrical and electronic calculators a century later. Unfortunately he didn't build any of his Analytical Engines either, although he did realize during his years of work on their designs (he produced a variety of plans and notes) that he could simplify his Difference Engine. So he sat down and designed Difference Engine No. 2, although this too wasn't built in his lifetime. Babbage was an example of someone who too easily moved on to the next project before completing the first, and in this respect also he was something of a forerunner to some of the twentieth-century pioneers.

For many years it was generally assumed that the main reason Babbage's "engines" were never completed was simply that they couldn't be built using the technology of the day. That was until a team led by Doron Swade, then curator of the computer collection at the Science Museum, was prompted by the Australian historian Allan Bromley to test this assumption by building a replica of one of them. They chose Difference Engine No. 2. Work started in the mid-1980s with the aim of having a functioning machine by December 1991, the 200th anniversary of Babbage's birth. They limited themselves to the materials and machining standards of the nineteenth century, and for time and cost reasons decided not to build the printing part of it. Thus began a six-year roller-coaster ride that involved fund-raising, corporate reorganization, seven-day weeks, a key contractor going bust, and much more, echoing Babbage's original trials and entertainingly recounted in Swade's book *The Difference Engine*. But the experiment worked and the result can be seen at the Science Museum today; a trip there is recommended for anyone who really wants to understand Babbage's engines. The printing and stereotyping apparatus was added later and was working by the spring of 2002. The exercise proved that Babbage's failure wasn't due to the limitations of nineteenth-century machining but rather to his dispute with his engineer, Joseph Clement, Babbage's inveterate tinkering with the design, and above all the fact that he just wasn't a very effective project manager.

Ever since Babbage's time it has been assumed that he was driven solely by his search for error-free calculation, but recent research by Doron Swade suggests this is a myth. Rather, says Swade, Babbage saw the engines as a new technology of mathematics able to systematically solve complex equations and compute functions for which there was no analytical law. Babbage even predicted a new branch of mathematics, which came into being much later under the label "computational analysis." Swade ascribes the enduring myth to Babbage's friend and advocate Dionysius Lardner, who made a living as a writer and traveling lecturer. Finding that his talk on the mathematical advances of Babbage's engines was too difficult for mass audiences to understand, he simplified the argument to concentrate solely on the avoidance of errors. When he wrote up his lecture in 1834 it became a standard reference paper and thus, as Swade concludes, "dumbing down has had a defining influence on our historical perception of Babbage's motives."

Babbage's Analytical Engine project is often cited as the first on which a woman programmer worked. Babbage's good friend (some say mistress) Ada Byron, Countess of Lovelace, was enchanted by his machines and particularly the mathematics involved and wrote a famous article, "Sketch of the Analytical Engine." This mixed her translation of a Swiss engineer's description with her own copious notes, including details of programming examples that most people have attributed to her. However, Doron Swade concludes that Byron's role in Babbage's work has been both exaggerated and distorted down the years, and describes her as a precocious novice where mathematics was concerned. Another noted Babbage historian, Bruce Collier, commented tartly, "I guess someone has to be the most overrated figure in the history of computing," but Swade does not go that far, arguing that Byron at least deserves credit for

> her unique understanding and insight into the potential of the computer particularly in areas beyond the confines of mathematics. She wrote of the Analytical Engine manipulating symbols representing entities other than quantity and the extension of the concept of a computer beyond number is not found anywhere else in contemporary commentary, and specifically not in Babbage's writing. So I would say that Byron deserves to be celebrated as the first person to see beyond Babbage, in a visionary and even prophetic way, the potential of a universal computing machine for application outside calculation.

Byron's rise to iconic status as the first "programmer" became unstoppable in the mid-twentieth century and was crowned in the 1970s by the naming of the computer language "Ada" in her honor.

Toward the end of the nineteenth century came another

important step on the road to the modern computer. The US Constitution stipulates that a census must be held every ten years in every state for the purpose of calculating the numbers in the House of Representatives. The first census was a simple head count, held in 1790 when the population was under 4 million, and took just nine months. By the late nineteenth century much more information was being gathered about each person, the population was approaching 50 million, and the analysis was taking seven years. Each census took longer than the one before and the government feared it would still be adding up the 1890 census returns by the time the 1900 census took place.

Enter the first revolution in office calculating machines, the mechanical tabulator. This was a set of equipment devised by the inventor Herman Hollerith. It enabled an operator to punch holes in a card corresponding to the data gathered about the individual (one punched card per person). An electrical machine read the card and indicated which of a large number of compartments to place the card in. Using these machines, the human counters could collate and analyze the figures in a fraction of the time it had taken to do everything by hand. The 1890 census results were available in just two years, saving the Census Bureau $5 million (equivalent to $100 million today).

Many companies had similar requirements for analyzing large amounts of data, so this early success started a huge industry in tabulating machines and development was rapid. The 1900 census was tabulated in a mere six weeks, despite a further 50 percent increase in population since 1880. Hollerith's firm merged with two rivals to become the Computing-Tabulating-Recording Company, and in 1924 it was renamed International Business Machines— soon to be known simply as IBM. Another early rival to Hollerith in the manufacture of tabulating machinery was James Powers, whose company in turn merged with others into the Remington

Rand Corporation in 1927 (Remington had become famous as the maker of the first commercial typewriter, while Rand made the Kardex card index systems). IBM came to dominate the tabulating-machine market, with Remington Rand taking much of what was left, and both companies would have important roles in the computer age.

There was one further period of innovation before the computer age really began. It started with the invention by Vannevar Bush of the Differential Analyzer, which he unveiled in 1930 at the Massachusetts Institute of Technology. Several more Differential Analyzers were built in Britain during the middle years of the decade and most of these were made largely from Meccano, the legendary boy's construction kit, but they were not toys and were capable of a surprising degree of accuracy. None of these British machines, however, matched Vannevar Bush's second Analyzer, unveiled in 1935, a 100-ton monster with around 2,000 valves (or "tubes"), almost as many relays, and 150 electric motors. Its calculations were controlled by instructions on paper tape and, although the computations were purely mechanical (the electronic components just controlled the movements of the mechanical ones), it was another significant step, not least for its successful use of so many valves.

The Differential Analyzers were of great importance in the 1930s, enabling the solution in a matter of hours or even minutes of complex differential equations that could take teams of human mathematicians weeks to solve. These were not necessarily obscure mathematical exercises, as differential equations could be used to model weather systems, describe the trajectory of a shell fired from a gun, or calculate the rate of erosion of river banks. They would later find many uses in war-related applications and influence the thinking of several of the early British, American, and Australian computer pioneers.

During the same decade a German engineer, Konrad Zuse, was working on one of the first electromechanical computers, the Z1. Born in 1910, Zuse was still a student when he started thinking about better calculating machines based on three logical principles: program control, the binary system, and floating-point arithmetic.

Zuse later justifiably claimed that "today these concepts are taken for granted but at the time this was new ground for computing." He started active design in 1934, "working independently and without knowledge of other developments going on around me," and within two years he finished the "logical plan." It took two more years to construct his machine before it started working in 1938. His son Horst Zuse claims it was "the first freely programmable binary-based machine in the world." As usual the carefully framed definition is crucial to the claim, but the Z1 was certainly a remarkable device. It had a memory of 64 words (each word containing 22 "bits," or "binary digits"). It had a high-performance adder and was capable of floating-point arithmetic, and hence it was able to handle very small or very large numbers with precision. It had a control unit, and the whole machine was programmed by instructions on paper tape. Perhaps the most ingenious parts were the mechanical "yes/no" modules used in both the memory and the arithmetic unit, where Zuse designed a system of sliding pins in a grid of movable thin metal sheets. This was analogous to the modern electronic memory that consists of a large number of bits in a "grid" of rows and columns, where every bit can be individually "addressed" by row number and grid number and then written or read.

Clever though the Z1 was, Zuse found it insufficiently reliable and went on to build the Z2, which used a similar memory but had 800 old telephone relays in the arithmetic unit. That convinced him that electrical relays were reliable enough and he went on to build the most impressive of the series, the Z3, which used relays throughout. Started in 1939 and completed in 1941, the Z3, Horst Zuse claimed, was "the first reliable, freely programmable, working computer in the world based on a floating-point number and switching system." It was big, at five meters long, two meters high and almost a meter deep, though not on the scale of the American "Harvard Mark 1" electrical relay computer, which it pre-dated by several years. Konrad Zuse claimed his Z3 could "calculate all mathematical problems" and was even capable of playing chess, though that was never demonstrated. He did, however, develop a sophisticated programming language for the machine, which he called "plankalkul."

Unfortunately for Zuse, all three machines and most of the drawings were destroyed in air raids that flattened Zuse Apparatebau, the company he had formed in 1940 in Berlin to build them. During the later war years he worked on the Z4, but relentless bombing of the German capital prevented its completion, and in 1945 he fled with the machine to Bavaria, where he hid it in a barn. For two years daily survival amid the shortages of occupied Germany was his main concern, but by 1947 he had the Z4 working again, though mains power was intermittent and he had to fashion spare parts from discarded tin cans. Eventually he persuaded the Swiss Federal Institute of Technology in Zurich to buy the machine and in 1950 he delivered it to the institute, where the scientists were astonished at the mechanical memory (he had reverted to the thin metal sheets and pins of the Z1 and Z2). It was reliable enough to leave running unattended overnight and Zuse once said that "the rattling of the pins and relays was the only interesting thing about Zurich's nightlife."

The Z4 did some useful work in Switzerland but, ultimately, half a second to perform a simple addition and six seconds to compute a single division were far too slow to compete with electronic methods of calculation. It was the end of an era rather than the beginning. Zuse had moreover missed an earlier chance to take the electronic route. Back in 1936 he was lobbied by his friend Helmut Schreyer to use valves which "could switch a million times faster than elements burdened with mechanical and inductive inertia." However, suitable circuits did not exist then and Schreyer's own colleagues doubted a machine with thousands of valves would work reliably. Yet before the end of the 1930s American and British pioneers were choosing the electronic option and in a few years would prove the doubters wrong.

CHAPTER 1

FROM ABC TO ENIAC

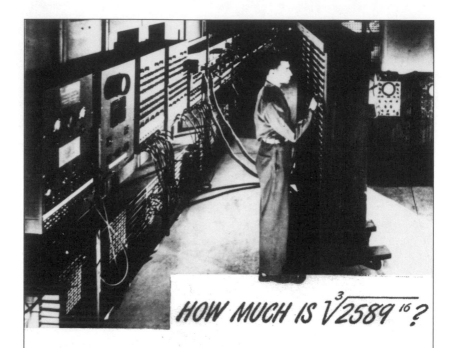

HOW MUCH IS $\sqrt[3]{2589}^{16}$?

The Army's ENIAC can give you the answer in a fraction of a second!

Think that's a stumper? You should see *some* of the Eniac's problems! Brain twisters that if put to paper would run off this page and feet beyond ... addition, subtraction, multiplication, division --- square root, cube root, any root. Solved by an incredibly complex system of circuits operating 18,000 electronic tubes and tipping the scales at 30 tons!

The Eniac is symbolic of many amazing Army devices with a brilliant future for you! The new Regular Army needs men with aptitude for scientific work, and as one of the first trained in the post-war era, you stand to get in on the ground floor of important jobs which

have never before existed. You'll find that an Army career pays off.

The most attractive fields are filling quickly. Get into the swim while the getting's good! $1\frac{1}{2}$, 2 and 3 year enlistments are open in the Regular Army to ambitious young men 18 to 34 (17 with parents' consent) who are otherwise qualified. If you enlist for 3 years, you may choose your own branch of the service, of those still open. Get full details at your nearest Army Recruiting Station.

YOUR REGULAR ARMY SERVES THE NATION AND MANKIND IN WAR AND PEACE

The US Army was quick to realize its need for electronic engineers and computer operators and exploited the recruiting potential of the ENIAC to the full (US Army photo and poster).

One winter evening in 1937 a professor from Iowa State College went for a drive along the open roads across the eastern half of the state into neighboring Illinois. Prohibition was still in force in some states, including Iowa, but if you were in a well-paid job then life was good and you could afford to drive your V8-powered Ford almost 200 miles along deserted highways to find a bar. The speeding professor's long drive that night was a way of working out his frustration with the current crop of mechanical calculators: they were slow, inaccurate, and not really up to solving the great mathematical problems he was interested in. He had been groping towards an idea for a more advanced machine that would calculate large systems of simultaneous equations and spit out the answer speedily and precisely, but "nothing was happening." If it seemed like a long way to go for a drink it turned out to be well worthwhile.

John Atanasoff wasn't everyone's idea of an academic in the 1930s. Born in 1903 and by then professor of mathematics and physics, he loved fast cars and whiskey and didn't mind mixing the two. His high-speed drive ended at a roadside tavern, where he downed a couple of shots before the basic components of a computer came to him in a rush and he hurriedly noted them on the back of a table napkin. Much later he was to say, in a story he loved repeating and one that may well have gained in the telling, "It was an evening of Scotch and 100 mph car rides when the concept

3

came for an electronically operated machine that would use base-2 [binary] numbers instead of the traditional base-10 [decimal], condensers for memory and a regenerative process to preclude loss of memory from electrical failure."

Safely back at Iowa State College, in the city of Ames, he began to sketch out the idea in more detail. While sitting in the bar he had only imagined a "black box" to do the arithmetic. Now he had to work out what would be in the black box. He chose to use logic circuits rather than simply mimicking the action of mechanical calculators, a far-sighted choice, as was his decision to use binary arithmetic instead of decimal. But he was not entirely alone in that decision. Quite independently and on the other side of the Atlantic, Alan Turing was developing his own binary logic device, an encoding machine, though this used mechanical relays rather than electronics.

For a working memory Atanasoff designed a rotating drum with capacitors on it. Each capacitor could be charged positively to represent a "1" or negatively to represent a "0." Because the charge would slowly leak away, Atanasoff designed a circuit to keep regenerating the charge on each capacitor (he called it "jogging"). In this he anticipated "dynamic RAM" (random-access memory) in the modern PC, which still uses the same principle. These and other details of the design occupied much of his spare time for over a year, but by March 1939 he was ready to apply for a grant to build the machine. The college award of $650 was generous for the time by academic standards, and it covered an assistant and materials. All he needed now was to find the assistant and somewhere to work.

Here Atanasoff was both wise and fortunate, realizing his own skills in mathematics and physics needed to be complemented by an assistant with good electronic and mechanical skills. A colleague recommended Clifford Berry, an outstanding student about to graduate and go on to a master's degree, and this was an inspired choice. Together they commandeered some space in the largely unused basement below the physics building and by October 1939, to the amazement of their colleagues, they demonstrated a simple prototype. It could add and subtract and the memory worked. It earned further grants of over $800 to build a full-sized model.

Work on the machine continued through 1940 though it was still a spare-time occupation for both men. They began to think about patenting it, and in August they wrote "Computing Machine for the Solution of Large Systems of Linear Algebraic Equations," a description of what was then known simply as "Atanasoff's machine." Only much later was it dubbed the Atanasoff Berry Computer, or ABC for short. The last part of their report was a request for a further grant of $5,330 from the Iowa State College Research Corporation to complete the machine and develop it further. The report included an explanation of how the computer could be used to solve a variety of problems in physics and applied mathematics. The grant was approved in March 1941 and the college appointed a lawyer to patent the machine. After some wrangling, Atanasoff agreed to split the patent income 50/50 with the college (which had initially suggested 90/10 in its favor). Unusually Atanasoff insisted that Berry would also have a share of his half, something unheard of for a graduate student.

In its finished form the ABC was as big as a fridge, weighed a third of a ton and used over 300 tubes, rather more than most electronic devices of the time, and it took 15 seconds to complete an arithmetical calculation. In addition to its binary logic and regenerative memory, it employed another innovation. This was a system that recorded the results of intermediate calculations on special cards, so they could be read back later in a computation. Mechanical punched cards would be too slow and cumbersome, so Atanasoff developed a system using another type of card that could be "written to" by electrical sparks instead of punching holes. This proved to be the Achilles heel of the machine. Very rarely, perhaps once in 100,000 operations, a figure would be written or read incorrectly, invalidating the whole calculation. This was a major problem, as solving large numbers of simultaneous equations involved many hundreds of thousands of operations (see Appendix B), so errors were inevitable.

It wasn't an insoluble problem but as Atanasoff and Berry worked on it through 1941, still in their spare time, a much bigger problem was just around the corner. On December 7, Japan attacked Pearl Harbor and World War II became truly global. Atanasoff had another project, a defense-related one, that was now urgent. In May 1942 Cliff Berry married Atanasoff's secretary and they moved to California, where Berry took a job in the defense industry. Atanasoff himself was then recruited to a wartime job at the Naval Ordnance Laboratory in Washington, DC. He left the ABC in the basement of Iowa State College, confident that the patent application was under way and that he would return to it when the war was over.

Unfortunately, for reasons that are unclear, the patent application was never followed through. After the war John Atanasoff continued his work in the Naval Ordnance Laboratory and declined an offer to return to Iowa State College as head of the physics department, an offer that would have meant he could have revived his computer project. Instead the ABC languished in the basement of the university and in 1948 the new head of department, believing it would never be used again and needing the storage space, ordered it to be dismantled. Its importance wasn't recognized, not even, it seems, by Atanasoff, and if it hadn't been for its influence on the creators of a much more famous computer, it would almost certainly have been forgotten forever.

Those creators were John Mauchly and Presper Eckert. Mauchly was one of the first people to entertain serious thoughts about creating an electronic computer. Art Gehring, who worked for him in later years, recalls "a professor of physics whose main interest in life was to build a machine that would predict the weather—this was his real passion—and he was playing around with electronic circuitry back in the 1930s, when physicists were coming up with differential equations that could be used to describe the weather." But these differential equations were difficult to solve and above all very time-consuming. If the calculations for the next 24 hours' weather take three weeks to carry out, then even the most perfect prediction is of no practical use. Some sort of machine was necessary to speed up the process, and on December 26, 1940, Mauchly met John Atanasoff for the first time.

By then John Mauchly had been professor of physics at Ursinus College in Philadelphia for eight years and that night he gave a lecture on his use of a "harmonic analyzer" in weather forecasting. It was more than a year since Atanasoff and Berry had demonstrated the first version of the ABC to colleagues at Iowa Stage College and they had developed it considerably during 1940. After Mauchly's lecture Atanasoff introduced himself and the two talked at some length about the ABC. It should be noted that the harmonic analyzer that Mauchly was lecturing about was an analog machine, like the differential analyzer, not a digital one, like Atanasoff's. An analog device operates continuously, like the mercury column that rises and falls in a traditional glass thermometer; in contrast a digital device operates in steps, like the numbers incrementing on the display of a modern (digital) thermometer. Digital measurements are more easily and accurately handled by electronic computers and digital operation was to prove fundamental to the development of the modern computer. So, *if* this meeting with Atanasoff was Mauchly's first introduction to the idea of digital computing, then it was a significant one. The extent to which Mauchly was influenced by Atanasoff's ideas and the ABC would be tested in court many years later.

What is not in doubt is that by 1941 Mauchly was very interested in electronic calculation as a way of achieving the speed necessary for weather forecasting. Ursinus College, however, wasn't noted for its electrical engineering, so a letter that landed on his desk in May was just the opportunity he was looking for. It was from the University of Pennsylvania's Moore School of Electrical Engineering, which was noted for its expertise in the emerging field of electronics, inviting him to nominate students for a ten-week summer school in the subject. The letter explained that the course was "in connection with the Defense Program" and was aimed at mathematics and physics students because there was a "serious, if not critical, lack of trained engineers available to industry to effectuate the industrial mobilization required in the time available." It was six months before the attack on Pearl Harbor that drew America into World War II, and the idea was to retrain mathematicians and physicists as electrical and electronics engineers. John Mauchly decided to enroll himself in the course.

Before the course began, Mauchly took up an invitation to visit Atanasoff in Iowa, staying with him for several days in June to see how he had put the theory of digital computing into practice. Some theoretical principles of electronic calculation were already becoming established. By using vacuum tubes as switches, a clever engineer could design a circuit that would add two simple numbers together. From that building block, he or she could develop more complex addition and subtraction and hence perform multiplication and division. That in turn would allow more sophisticated mathematics, including the vital solution of differential equations. The real challenges were practical: to actually build a machine that would work. But Atanasoff had done it, at least on a small scale and for the simpler task of solving simultaneous equations.

Shortly after returning from Iowa, John Mauchly went to the Moore School to take up his place on the "Defense Program" course and spend ten weeks learning about electronics. One of his teachers was a young man whose brilliance as a student had won him an instructor's post on graduation: this was Presper Eckert. The course turned out to be straightforward for Mauchly. Indeed, he so impressed the university that he was offered an academic position there and he never returned to Ursinus College. He also had many after-hours discussions with Eckert about the possibility of an electronic computer and this was the start of the partnership between the two men, one an experienced professor in his mid-thirties, the other a bright young graduate of 22.

What John Mauchly had in mind was a much more complicated machine than the ABC, using what was then regarded as an enormous number of components, bearing in mind that electronic devices of the time, such as wirelesses, used only a handful of tubes. Even the 300 or so used in the ABC was pushing the envelope (the electromechanical Bush Differential Analyzer was already using some 2,000 tubes, though not for the actual calculations). But Mauchly was working towards a device with 18,000 tubes. Even if such an electronic calculator were built, would it ever complete a calculation between tube failures?

That wasn't the only practical consideration, not by a long way. The investment required would be huge, vastly more than Atanasoff's few thousand dollars. During his first year at the University of Pennsylvania, Mauchly struggled to get anyone interested in spending that kind of money on the promise of better weather forecasts. However, America's entry into World War II opened up another possibility.

By the middle of 1942 the Moore School housed a room full of women mathematicians doing vital war work. One of them was Jean "Betty" Bartik, who recalls that female mathematicians were a rarity then. "I was the only woman mathematics major in my college. The jobs market was a big difficulty and we thought all we could do was teach school, and I definitely did not want to teach school. My calculus teacher knew that and she received this recruiting notice from Aberdeen Proving Ground looking for women math majors and she gave it to me. I applied for the job as a 'computor'—my title was 'computor'!" Indeed all the women in that room were "computors": the label applied to female mathematicians long before it became synonymous with electronic boxes. It wasn't even a very high-status name. It was a semi-professional grade and most "computors" were women, while men were "mathematicians," professionals. But at least it wasn't teaching.

The job the women were doing at the Moore School was calculating ballistic tables for the military. New weapons were being developed at a great rate for the war effort, and many were tested at the Army's Aberdeen Proving Ground, 80 miles away in Maryland. For each new gun the armed forces needed a table showing the distance a shell would travel for every firing angle between 5 degrees and near vertical. So they had to compute a lot of firing tables, and that was the reason for the job ad Betty Bartik saw. "They needed to calculate the trajectories of bullets or shells. Each trajectory for one firing of one shell took about 40 hours—and each table had hundreds of these trajectories—so naturally they needed a lot of people to calculate these. I found it interesting but it wasn't creative; I was just doing the calculations." A single firing table took this whole team several weeks to complete.

Also working at the Moore School was Kathleen "Kay" McNulty, who was to become a good friend of Bartik's. She had been a bright child who also majored in mathematics at college but, like Bartik, she had found few openings for her skills. "Insurance companies had a lot of actuarial work. However, they didn't hire women, because by the time they'd trained a woman to be an actuary she was ready to get married and leave. That was the big thing before the war—once you got married you no longer worked, you just stayed home. I started taking various business courses, thinking I might end up in a bank or something; anything I could get, I took." So she too had responded to the army's call for women mathematicians: "When we got there they asked us if we'd done differential equations and calculus and so on. We had taken every course going, so we said yes, and they hired us and told us to report to the University of Pennsylvania. They didn't say what we were being hired for but we were going to be civilian employees of the Aberdeen Proving Ground."

When Kay McNulty joined the female computors at the university, she soon found there was already a way of reducing that figure of 40 hours to calculate a single firing, as that calculation was made using a desk calculator. "A differential analyzer would do the same calculation in three-quarters of an hour, and the university had one of only five in the world, of which theirs was the largest. It was about 20 feet long, a very complicated machine but purely mechanical, just made up of shafts and gears and motors and things like that. They said they needed people to operate this differential analyzer and they assigned two of us to it. We did that for the whole war; we worked there in the basement of the Moore School on this tremendous machine eight hours a day, six days a week, no vacations, just July 4 and Christmas. Once we knew what the equations were that we had to use, it was just routine."

That differential analyzer was similar to the device invented by Vannevar Bush in the 1930s and copied in other parts of the USA and abroad. This version had been built in the Moore School mainly for academic and civil applications, but now it was being put to work in support of the war effort. Seeing it in operation was another part of the jigsaw for John Mauchly. Kay McNulty recalls that "after I was there for a while Mauchly got interested in what was going on here, what were all these people being hired for? So he came down, he saw what was going on and right away he thought, 'Boy, that's just what I want—somebody to build this electronic computer—it has to be the army who needs all this computing done!' As early as August of 1942 Mauchly wrote a note to the other professors saying that he was proposing building an electronic computer that would do the work that the differential analyzer did. Mauchly figured out that if he could get a computer that would operate at 100,000 pulses/second—and a pulse would represent an addition or something—he could do a whole firing calculation in 10 minutes or even less." This was the basic idea for the ENIAC, and Kay McNulty regards the differential analyzer as "the beginning of the ENIAC."

There is some confusion regarding precisely what "ENIAC" stood for. Many people, including some involved in the original project, say that it was "Electronic Numerical Integrator and Calculator." Others say it was not "Calculator" but "Computer"— although the term then meant a female mathematician. Maybe that was the point, that this was an electronic replacement for the female "computor." Some argue it originally stood for "Electronic Numerical Integrator Analyzer and Calculator" (or Computer) and given that Mauchly initially proposed it to the army as an electronic differential analyzer that too is plausible. However, reproductions of Moore School reports dated 1944 and 1946 clearly show that it then stood for "Electronic Numerical Integrator and Computer," although a diagram inside one of them shows the spelling as "Computor" so even that detail is open to question.

Whatever it stood for, the idea was not received with enthusiasm to begin with, as Kay McNulty recalls: "They were all too busy to work on it and in 1942 everybody thought the war would be over in a year, so they just pooh-poohed it. But everything just kept getting worse and worse, and more and bigger and better guns were being made, so they needed more and more firing tables. Pretty soon they had up to a hundred women there computing these trajectories." As a result, Mauchly's idea began to be taken more seriously and Dr. Nathan Ensmenger, a technology historian and sociologist at the University of Pennsylvania, agrees this need was crucial to getting the project funded: "The US government was very willing to invest in machinery for computation during World War II and it had this particular problem that it needed ENIAC to solve, the construction of ballistics tables. So they had a very practical purpose in mind, they were willing to spend the money, and it's questionable whether any group other than the US government could have produced such an expensive machine. It was an untried technology and it would have been quite a risk even for IBM; it probably would not have been cost-effective and the payoff was very uncertain. But the government was willing to do it and it had the resources."

Another important factor in the growing acceptance of John Mauchly's idea was the contribution of Presper Eckert, who was as fine an electrical engineer as Mauchly was a mathematician and physicist. It was an almost precise copy of the Atanasoff–Berry relationship. Kay McNulty says, "Eckert was without any doubt whatsoever the most brilliant student they had. He knew more about electronics than any of the professors there—he just was a natural genius, and even at the age of 18 he already had a patent for some things. You made a proposal to him and he would come up with an idea and he would make it work. So, of course, Mauchly, thinking that they would build this machine, went to Eckert and said, 'What do you think, can you do it?' Eckert replied, 'Yes, if you're very, very careful about how you use these things [tubes].' He was sure that they could build some sort of a machine that could actually count at the rate of 100,000 pulses per second."

So Eckert and Mauchly started the design and McNulty says they based it on the way she and her colleagues tackled the problem: "If you were doing this job by hand you had a large sheet of paper, which was about 28 inches by 20 inches. It was divided into about 14 different columns and each one of these columns had some mathematical part. And for every tenth of a second you calculated where the shell would go, going through all these 20 steps as it went along. So when they decided to build a computer, they decided, well, we need these 20 spots where we can put the calculations and that was the way they started out, they built these 20 'accumulators,' which were named exactly as we had headed up these big sheets of paper!"

The fundamental principle of each of the ENIAC's accummu-lators was quite straightforward. Mauchly used the analogy of mechanical calculators and Eckert refined a device already in use in the 1930s, the electronic "ring counter" that was used in the measurement of frequencies (of radio signals, for example). At its simplest, the ring counter consisted of ten tubes in a circle, each representing one of the digits 0, 1, 2, 3, 4, 5, 6, 7, 8, 9. Only one tube could be "on" at a time and initially this would be 0. Each time a pulse came into the counter that tube would go off and the next one would come on. So to represent the number "3" you would input three pulses and tube number 3 would switch on. To add 6 to 3 you would put in another 6 pulses—each one would move the switched tube round another step until eventually tube number 9 lit up (for good electronic reasons each tube was actually a "flip-flop," a pair of tubes where if one is "on" the other will be "off").

What happens if you want to add say another 5 to the total? The five pulses would advance round the ring until tube number 4 was lit up and as it passed through 0 an extra pulse would be sent to the next ring counter, the one counting "tens." Hence 1 ten, plus 4 units, representing 14. This of course was decimal arithmetic, making it a rare design among electronic computers; both the earlier ABC and nearly all future computers would use the "binary" system, where numbers are represented by 0s and 1s only.

Each of these large "accumulators" had ten "ring counters," so one accumulator could store any number between zero (represented as 0,000,000,000) and 9,999,999,999. If you could do addition, then you could reverse the process and do subtraction; you could do multiplication simply by repeated addition and counting the number of repetitions, and division was just a matter of repeated subtractions. Do these four basic mathematical operations often enough, in the right order, using all 20 accumulators and you could solve the differential equations that defined the path of a shell from gun to target. Although such complex calculations required many millions of pulses through the system, it wouldn't take very long operating at 100,000 pulses a second. Mauchly reckoned about 5 minutes, an estimate that was to prove rather conservative.

All he needed now was some enthusiastic and influential support, and that came from Captain Herman Goldstine, a mathematics Ph.D. from the Army Ordnance Corps. He had been posted from the Ballistic Research Laboratory at the Aberdeen Proving Ground to supervise the work of the group of female "computors" at the Moore School. He was in effect the army's representative at the university, but he was also a powerful advocate for the university within the Army Ordnance Corps. The Moore School itself was run by Professor John Brainerd and these two men backed Eckert and Mauchly's scheme strongly, recruiting the support of leading figures in the Army's Ballistic Research Laboratory. On April 8, 1943, they sent a formal proposal to the chief of ordnance, and the first contract was signed on June 5. Eckert and Mauchly would be the lead designers under the supervision of Brainerd, with Goldstine the army liaison officer. The contract authorized six months of research and development for an "electronic numerical integrator and computer" at a cost of $61,700. Once under way the project was unstoppable and the eventual cost to the army totaled almost half a million dollars in payments to the Moore School alone.

The army got a lot for its money. The ENIAC was a monster by any standards, occupying a room 50 feet by 30, with its 18,000 tubes, 70,000 resistors, etc. The heat generated was enormous and the cooling system alone weighed several tons.

While Mauchly conceived the overall idea and the mathematical capabilities that each stage of the machine had to have, Eckert was the one who put it all into practice. Art Gehring reckons Eckert "was a genius when it came to designing circuits. There were a lot of people who said that you can't get that many vacuum tubes to operate at the same time without making an error, you just can't do it. But he did it, his design was so reliable it was one of the big things that made the ENIAC work. He was able to design it so it would tolerate quite a bit of variability in the power input and the vacuum tubes and it would still work. Vacuum tubes up to that time were analog devices, so digital use [where they must be either "on" or "off"] was something brand new."

Not everyone was so impressed by their virtuosity at the time, Kay McNulty for one: "One evening Betty Alice Snyder [another 'computor'] and I were working on the differential analyzer and both Eckert and Mauchly came down. They were all excited and they said, 'You have to come up and see what we have accomplished.' This was about a year into the construction. Outside the room where the ENIAC was being built there was another small room called a 'high-voltage lab' and in there Eckert and Mauchly had built the first two accumulators, which were about eight feet tall and two feet wide, and a power supply. So they took us up and Eckert punched up numbers so a 5 appeared on one of the little lights on the panel. And he said, 'Now watch this,' and by just using the punch again it appeared as if this 5 jumped over to the other panel and became 5,000. And he said, 'We just multiplied that 5 by 1,000,' and Betty Alice Snyder and I just looked at each other and said, 'You mean you used all that equipment just to multiply by 1,000?'"

That simple demonstration was more significant than it appeared. Quite simply it showed that the ENIAC would work. If two accumulators worked together, then in principle you could just keep adding more accumulators until you had enough processing power to solve your problem. However, there was a lot more practical work to be done before the whole machine could be commissioned. In any event, it was never used in anger before the end of the war, but fortunately for Eckert and Mauchly—and the development of computing—the US Army didn't abandon the project. By then it was nearing completion and there were plenty of other demands for its computational abilities.

With Victory in Europe declared and Japan close to defeat, the demand for firing tables slumped. But as one door closed, Kay McNulty found another one opening: "As soon as the war ended, Aberdeen sent out notice that all the women who were hired in Philadelphia were to be laid off. However they said if you wanted to work on the ENIAC, you could apply to train as a programmer. I was one of the five chosen and we were sent down to Aberdeen, Maryland, to learn all about IBM [punched-card] equipment because the input and the output of the ENIAC was to be on IBM machines. We went down there for ten weeks and when we came back to Philadelphia the ENIAC was still not finished. That was a few days before VJ Day. I remember that because there was such a great celebration when Japan surrendered."

Betty Bartik was another of this select crew of five. After just three months as a "computor" she too saw the advertisement for

ENIAC programmers and applied, even though she didn't really know what one was. "They said it was going to calculate trajectories. One of the first things they asked me was, 'What do you think of electricity?' I said I know that $E = I \times R$ and he said, 'No, no, I'm not interested in that. Are you *afraid* of it?' He wanted to make sure I wasn't afraid of plugging in electrical cables." Bartik had little experience of job interviews and just barely made it into the five selected: "I was the second alternate. During the war, housing was terrible but the first choice for the job had a nice apartment in west Philadelphia and she knew that Aberdeen was a hell-hole, so she decided not to take the job. The first alternate was away on vacation. She also knew Aberdeen was a hell-hole, so she decided to stay on vacation, and that's how I became an ENIAC programmer." On such slim chances are life-changing decisions made: "I was glad I did. My God, I felt like I died and went to heaven. Working with Pres [Eckert] and John and these people, it was fun. People think it's hard to work for a genius; it's not, they're the easiest people in the world to work for 'cause there are no dumb questions. They would think with their mouths open so they were fabulous teachers. Pres used to bring something up with me and I would say, 'Pres, I don't know anything about that,' and he would say, 'That's all right, I'll explain it to you.' His whole day was spent going from one group to another, talking to them. When he was thinking at a snail's pace I was galloping along to keep up."

When they returned to the Moore School from their ten-week course, it was a long time before they were allowed even to see the ENIAC, so they had to work out how to program it from the circuit diagrams. Not only was programming a very new skill, there was no internal memory to store the program, so the steps were defined by the way the accumulators were wired together, using a large number of pluggable cables and switch panels. It wasn't necessary to rewire the machine for every single calculation as certain values could be set on the switches. So, for example, to create a firing table for a particular gun it might take days to rewire the machine, but then each trajectory would need only a change of input data and the whole table could be calculated in a matter of hours.

One advantage of having all these separate accumulators was that they could operate at the same time—in other words in parallel. So while one black box was calculating one part of a complicated set of differential equations, for example, another would be working out a different part and the results were combined later in the process. But that was another reason why it was so difficult to program, as all these simultaneous calculations had to be coordinated.

At last, late in November 1945, the great day came. "We were told the ENIAC was ready to be operated. Well, this was a joyous moment for us and we went down to see it," remembers Kay McNulty, "but what a sight it was. It was 80 feet long, it went around three sides of a room, it had 40 separate panels, each one 2 feet wide and about 8 feet high and everything was black. It was very gloomy-looking altogether. It didn't look like we expected, because we'd seen all these IBM machines down at Aberdeen which were beautiful stainless steel, so we didn't expect these to be nothing but black. That is what I remember most about it, the blackness. It was really weird. I remember the sound, because in order to carry off all the heat that was generated by 18,000 tubes they had to put this huge exhaust system in the ceiling."

They weren't just there to look at the machine though; it had a real problem to solve and they were to program it by wiring up the interconnecting cables and setting all the thousands of switches. It was a top-secret test, but Kay McNulty and her fellow programmers had a good idea what was going on. "Dr. Goldstine, who had been a first lieutenant originally and was now a captain, was there with two young physicists from Los Alamos. They had come east and had learned in the three months since the atom bomb was dropped all about how to program the ENIAC and they had a problem, which was the feasibility of the hydrogen bomb." Though the world war was over, the Cold War was just beginning and the team that had built the atom bomb was now looking at the next stage, the device that would become known as the H-bomb. This uses the enormous heat, pressure, and radiation of an atom bomb to trigger the fusion of hydrogen, turning it into helium in a truly massive explosion. But the mathematics involved is immensely complicated, and only the ENIAC was capable of computing it in a reasonable time.

"We didn't know what it was they were working on," says McNulty, "but we knew they were from Los Alamos, so we had to assume it was something to do with nuclear fission. They had typed up little cards telling you how to set all the switches and the lines that carried the pulses. That was its trial run, the first time it had ever been tested, so there we were—'Go ahead, girls, plug it up!' which we did." Betty Bartik too remembers that test vividly: "Herman Goldstine was like the conductor of an orchestra as he read out the instructions. He was yelling out, 'Accumulator 1, switch 1,' and we were busily following these directions." This was the first full-scale operational run of the ENIAC; it was successful and it was an impressive demonstration of the machine's potential.

There were no celebrations after that first run, or if there were the programmers weren't invited. But Bartik remembers a different story when the machine was introduced to the public in February 1946: "That was when the excitement began because the press had all these idiotic articles about the 'thinking machines' and cybernetics. Scientists from all over the world began arriving, and then Pathé News, so it was really exciting. The big panels of the ENIAC at the top had little holes and the tips of the vacuum tubes showed through these matrices of little holes, so when you did a problem these lights went up and down and flashed. Then Pres and

John found they didn't show up on the cameras for Pathé News—so they got little tiny neon bulbs that they screwed on the ends of them so they lighted when the vacuum tubes lighted. Then they turned out all the other lights in the room and ran the trajectory calculation and it actually ran faster than it would have taken the shell to fly to its target. In fact in Hollywood, for many years after that, when you saw a computer, what you saw was the ENIAC with these lights flashing." Eckert and Mauchly later went a step further, cutting ping-pong balls in half, writing a digit on each one, and sticking them over the neons, so the results were displayed in easily readable form for the audience.

The interest was intense and worldwide. McNulty recalls they had an order for an ENIAC within 10 days—from Moscow. While that might seem rather improbable, the evidence is there in the public record. The order was not accepted.

Even as the ENIAC was taking shape, and long before its public debut, Eckert and Mauchly came to regard it as obsolete. Only one ENIAC was ever built. Its successor would use binary arithmetic instead of decimal, which would drastically reduce the number of components involved, and it would have an internal memory to store the program and the data, to make programming much quicker. It would be a true binary stored program computer.

They could also see commercial possibilities for their next design. In this respect the technology historian Nathan Ensmenger regards the ENIAC creators' wider vision as remarkable: "I think very few people in 1946 had any idea that computing would become a commercial activity. The people who first used the ENIAC to solve significant problems were the US military and the Los Alamos laboratory, where nuclear weapons development was occurring, and they had a very different image of what a computer would be used for. What Eckert and Mauchly, and in particular John Mauchly, brought to this was a vision of a computer as a commercial instrument. It was a vision that was not shared by many people in the late 1940s and that made them very prophetic about the technology and where it was going."

Everything was in place for some of the most significant developments in computing history, accompanied by enough controversy, fallings-out, setbacks, smears, financial mismanagement, and even sudden death to form the basis of a thoroughly improbable soap opera.

The first falling-out was precipitated, probably unintentionally, by John von Neumann, one of the world's leading mathematicians, who had been a part-time consultant to the ENIAC team since 1944. Von Neumann was born in Hungary in 1903, published his first mathematical paper at the age of 17, and had established a worldwide reputation by his mid-twenties. He emigrated to America to become one of the original six professors of mathematics in the prestigious Institute for Advanced Study (IAS) when it was first established in Princeton, New Jersey, in 1930. During the war he was much in demand for a variety of military projects and the atom bomb work at Los Alamos in particular. It was a chance meeting with Herman Goldstine on a platform at Aberdeen railway station in Maryland that led to him joining the ENIAC project. It was his idea to bring the hydrogen bomb problem to the ENIAC and make that its first full-scale test.

Von Neumann's contributions seem to have been welcomed by the rest of the team until, on June 30, 1945, he published a paper entitled "First Draft of a Report on the EDVAC." The acronym stood for Electronic-Delay Variable Automatic Calculator and this was to be the successor to the ENIAC. Von Neumann's report set out the crucial features of input, processor, control, output, and memory. Moreover, the memory would be for both data and program—a computer with such a memory is called a "stored-program" computer. Importantly, that implies not only that the program instructions and the data share the same memory space, but that the computer can (at least in principle) build or modify its own program. Virtually all electronic computers since then have used "von Neumann architecture." Only von Neumann's name appeared on the report, although it drew on discussions about the design for the EDVAC that had been going on for months with Eckert, Mauchly, and other members of what had become quite a large team. Though nominally a "draft," he distributed the report widely and it became one of the most important documents in early computing. Arguments have raged ever since over how much credit von Neumann deserved for the definition that bears his name, and relations between Eckert and Mauchly on the one hand and von Neumann on the other deteriorated steadily after the publication of that report.

The problems multiplied when Eckert and Mauchly got into a dispute with the university over who owned the patent rights to the ENIAC. Kay McNulty says, "When the army first signed a contract for the ENIAC, the Moore School did not want to go to the expense of hiring patent lawyers. . . . I don't think they had much faith in this machine; they didn't really think it was going to work. So they told Eckert and Mauchly to take out the patents in their own names, and they would have to pay the lawyers. Eckert and Mauchly gave the University of Pennsylvania the right to use the patents for their own use, but they reserved for themselves the commercial use of the ENIAC—Mauchly could already see all the uses this could possibly have. Irven Travis, a professor who had been drafted into the navy, came back at the end of the war to take charge of the Moore School, and when he saw that Mauchly and Eckert had the patent rights to this thing, he exploded! He went to them and said, 'We want you to sign over all your patents to the Moore School,' they said no, and he gave them 15 days to sign or get out. So they got out—on a shoestring; they had no money and no contracts. Not only that but all the engineers had to sign this agreement or get out, so they too got out."

The breakdown in relations wasn't complete. Another professor at the Moore School, Carl Chambers, saw the damaging effect this loss of knowledge would have on the university and persuaded it to put on a summer school in 1946, much of it to be taught by Eckert and Mauchly. Large numbers of academics, engineers, and others from home and abroad were invited, a remarkable step given the military importance of their work and the absolute secrecy surrounding British and Soviet computing in that era. Kay McNulty says it was explicitly intended to "teach the world all about electronic computing. Eckert and Mauchly gave most of the lectures but all the other engineers who had worked on particular parts also spoke. People came in from the army, the navy, MIT, Harvard, and from every place to spread the doctrine of electronic computing. So that kept Eckert and Mauchly alive for that summer. This was giving away all their information but they had the patents and the government wasn't going to build commercial computers."

The Moore School lectures of summer 1946 were of enormous influence on the development of computing. They meant that many of the computer projects that started up in America and beyond grew out of Eckert and Mauchly's work. The summer school was evidence also that the departure of Eckert, Mauchly, and many of their team hadn't brought the university's computing effort to an end, although it did rather lose momentum.

Meanwhile the ENIAC was moved to a permanent home in the army's Ballistic Research Laboratory at the Aberdeen Proving Ground in Maryland. This was quite an undertaking according to Kay McNulty: "They had built the machine in a classroom, with windows, and one side was just a wall on to a courtyard. So they moved a truck into this courtyard, and tore down the wall and just moved the ENIAC right out where the wall had been. They told us to go on leave, because we hadn't had any during the four years of the war, and report back on December 1, 1946, to Aberdeen, where the machine would be installed. So we went on a trip round the United States for ten weeks, but the ENIAC spent practically the next year being reinstalled!" It was operational again in August 1947, churning out firing tables and every so often being re-programmed to solve other complicated military ballistic problems, including rocketry. Not only was the process much quicker than if calculated by a large roomful of human "computors," it could be carried out in much more detail and required fewer approximations, so it was more accurate. The main drawback was the system of programming by cable-plugging and this was tackled by introducing a system of removable and rewirable plug-boards. These could be programmed away from the ENIAC while it ran other problems and then installed as required.

The ENIAC was switched off for the last time at a quarter to midnight on October 2, 1955, ten years from Kay McNulty and Betty Bartik's first sight of the machine in action. There is a precise record of its operating time: for 80,223 hours, it processed 5,000 arithmetical operations per second.

While the ENIAC was being turned into a reasonably reliable workhorse, the Moore School and the US Army worked on its successor, the EDVAC. The first contract had been signed on April 12, 1946, for a preliminary model at a cost of $100,000. This would lead to the full-scale computer and again the total sum paid to the Moore School approached half a million dollars. A mercury delay-line memory was developed the same year and that became the internal memory that von Neumann's landmark paper had specified. But the loss of Eckert, Mauchly, and others meant that progress was slow and costs spiraled, in spite of the considerable institutional resources the army and the university could bring to bear.

A great deal of work was carried out on the project over the next few years, with the construction of intermediate models at the Moore School. Finally the full version was delivered to Aberdeen in August 1949. Although the logical design proved good, there were many problems with marginal circuit performance, perhaps a consequence of losing the electronic design skills of Eckert and others from the team. A further 18 months were lost in solving these problems and even limited operations did not begin until late 1951. Early the following year it was managing a rather unimpressive 15–20 hours of work per week. Ten years later it was operating a far more respectable 145 hours in each 168-hour week. Only one full EDVAC was built and it's fair to say it had little impact on computer development; in fact some otherwise authoritative histories erroneously but quite understandably say it was "never completed."

Although John von Neumann had his name on the EDVAC reports, he didn't stay with the project much longer than Eckert and Mauchly. After the war he returned to the IAS at Princeton full-time, having persuaded them to fund the development of their own EDVAC-type computer, and Hermann Goldstine went with him, having taken von Neumann's side in the disputes with Eckert and Mauchly. Unfortunately in his design for the IAS computer, von Neumann decided to use a device called the Selectron for the main memory that would store both program and data. This had been developed by the RCA electronics company and a prototype was presented by its inventor, Jan Rajhman, an RCA engineer, to participants in the Moore School summer course in 1946 as the answer to the problem of making a practical program-store. It was a large, specially designed vacuum tube with provision for writing, storing and reading 4,096 bits of data, and a production target of 200 was anticipated by the end of that year. However, two years later there was still no working production model, so von Neumann decided to switch to the newly successful Williams–Kilburn tube, developed in Britain. But their early lead had been lost and it was 1952 before the IAS machine became fully operational.

As the IAS project was jointly funded by a number of government agencies such as the Atomic Energy Commission and private companies such as IBM and RCA, its plans were widely distributed and a number of other computers derived from it, including Illinois University's ILLIAC, the Rand Corporation's JOHNNIAC (named after John von Neumann), and the MANIAC at Los Alamos. That so many early computer names ended in -IAC was an implicit acknowledgment of the debt they owed to Eckert and Mauchly's ENIAC.

The Selectron was eventually made to work in a less ambitious 256-bit mode and found a home in some of these other IAS-derived designs (the JOHNNIAC used 80 of them), but it wasn't one of the great successes of computer history.

The original IAS machine played an important part in the continuing work on developing the first hydrogen bomb. Von Neumann attended a number of atomic bomb tests and his premature death from cancer in 1957 was attributed to radiation exposure.

While the EDVAC found a useful niche in army ordnance work and the IAS spawned a family of important computers, Eckert and Mauchly's story after they left the Moore School was much more interesting and opened up a new chapter in the history of computing.

CHAPTER 2

———————

UNIVAC—SAVIOR OF THE CENSUS

Top: This early ad for the UNIVAC has the striking image of an office desk inside an electronic valve. It doesn't call the UNIVAC a "computer"—it is a "fact-troller," described as "the first universal electronic system designed for *both* management and science" (by permission of the Unisys Corporation).

Bottom: Jean "Betty" Bartik and Kay Mauchly Antonelli (née McNulty), pictured here in 2001, were two ENIAC programmers who contributed much to the UNIVAC project and in later years to the historical record (author's photographs).

When John Mauchly and Presper Eckert left the University of Pennsylvania in 1946, they decided to set up their own company in Philadelphia. The first thing they needed was premises. Eckert's father was in real estate and he found a clothing store on Walnut Street with two empty floors over it. They rented it and started the Electronic Control Company. They won some minor contracts to build small electronic devices and the engineers who had left the university with them joined the company, even working for nothing at first. Like Betty Bartik, Kay McNulty stayed with the ENIAC team for a while longer.

While the small contracts brought some money in, the real prize and the purpose of the new company were to build an advanced computer. This wasn't to be a one-of-a-kind like the EDVAC. They intended to sell it for a variety of military, scientific, and business applications, which is why they gave it the name UNIVAC, short for Universal Automatic Computer. Even before they'd come up with the name and were still calling it an "EDVAC-style machine," they went to the Census Bureau for support. Just as the Census Bureau had faced a crisis at the end of the nineteenth century, when manual counting was too slow to keep up with the rapidly multiplying population data, so again by the 1940s the census results were taking too many years to process. IBM's Hollerith-type tabulating machines were coming to the end of their useful life as a technology; more speed was needed and Eckert and Mauchly's

31

machine might just be the answer. The bureau couldn't fund the development work directly, so the money actually came from the National Bureau of Standards.

Unfortunately, while the two men clearly had some commercial acumen, they were in Kay McNulty's opinion "never businessmen" and they grossly underestimated both the cost and the timescale of their project. Their contracts with the National Bureau of Standards to supply the first "EDVAC-type machine" totaled under $300,000. Many years later it was reckoned that first machine had cost over $900,000, with much trauma along the way. The research phase was meant to take six months; it took a year. And so it went on.

They had big ideas from the start, says McNulty: "They decided that they had to take on IBM. Practically all the installations [tabulating machines] in the world were running on IBM cards in those days, so they had to go into an entirely different technology and do away with punched cards. So they set up a plating laboratory and started to try and manufacture some kind of magnetic phosphor-bronze tape that would carry the information into the computer and could take a lot of stopping and starting. This was all new and they had to teach themselves, but they had big dreams." The development of this special tape goes a long way to explaining the time it took to bring the UNIVAC to fruition, as well as the escalating costs. On the other hand, the UNIVAC was simpler in some ways than the ENIAC. Adopting the binary system of arithmetic that would become the norm in computing helped bring the tube count down from 18,000 to 5,200, the weight from 28 to 14 tons, and the power consumption from 174 to 125 kilowatts.

Eckert and Mauchly had many rivals, and one of the most influential politically was Howard Aiken, a professor of applied mathematics at Harvard University. He had been an important pioneer of electromechanical calculation and is most famous for his design of the Harvard Mark 1 during the early 1940s. This was a significant computer even though it performed its calculations using a mixture of electrical and mechanical components rather than electronics. Aiken allegedly asserted in 1948: "There will never be enough problems, enough work, for more than just one or two of these computers." If he really did say that, it might explain his efforts through the National Research Council to pressure the US Bureau of Standards into withdrawing financial support from Eckert and Mauchly. Since by that time Aiken had completed the Harvard Mark 2 (a faster version of the Mark 1, but still electro-mechanical), it seems he saw no need for any more computers than the two built to his designs.

That's not to say that Aiken's Mark 1 and 2 computers weren't important—they were. And if Aiken showed some meanness towards competitors, he deserves credit for encouraging a gifted young woman who would become another of the great names in computing history. Grace Hopper was one of the first recruits to Aiken's Bureau of Ordnance Computation project at Harvard; there she learned how to program the Mark 1 and became one of the first software engineers. She is credited with inventing the "compiler" (a program that converts English-language instructions into the computer's own machine codes) and a host of other software developments. Ironically, she left Harvard in 1949 and took her programming skills to Eckert and Mauchly's company, which had survived Aiken's attempted sabotage.

With rivals like Aiken and their former colleagues Goldstine and von Neumann, not to mention their own financial incompetence and optimistic timescales, it may fairly be assumed that the odds were somewhat against the UNIVAC ever being made. Eckert and Mauchly's financial savior arrived during 1948 in the unlikely form of the company that made gambling equipment for the horse-racing industry. Henry Straus was president of American Totalisator, a company he had created. Straus didn't want to take over; he was happy simply to invest a very large sum of money in return for a minority of shares in the Eckert–Mauchly Computer Corporation, as the Electronic Control Company had become in December of the previous year. Unfortunately, as it was to turn out, some of the investment was in the form of loans.

One of the other consequences of Straus's involvement was that some of his bright young men moved to Eckert and Mauchly's company. Max Kraus joined American Totalisator as a junior engineer straight out of university in the summer of 1945. When he transferred to Eckert–Mauchly in 1948, "the factory was a two-story loft without partitions or offices where I was. I was put to work under Brad Shepherd, who was the engineer, and we designed a memory test circuit. The UNIVAC had large mercury memory tanks that stored the information and the first thing we had to do was check them to see if the information really ran around this tank. Occasionally Pres Eckert would come round and see what we were doing. Brad was a great guy to work for and it was a community of people working together, with the same sort of spirit that was behind Silicon Valley. I remember the technician who worked with me liked opera and he had *Madame Butterfly* on his wire recorder and he would play it over and over all night. Even now when I hear the gunshot in *Madame Butterfly* I think back to Jack and his wire recorder. We were all young, all doing interesting things."

By this time the company was also working on a project for a simpler computer called the BINAC, or Binary Automatic Computer, the name highlighting the adoption of the binary system. It had won a contract from Northrop Aircraft and started work in October 1947, but again both price and timescale were hopelessly optimistic ($100,000 with delivery by May 1948), which is a pity, because if it had been ready on schedule it would have been the world's first stored-program electronic computer.

The BINAC was an unusual device, in effect two computers joined together, each checking the results of the other. It was commissioned to test a navigational system that would be used to guide the Snark nuclear missile and it eventually ran satisfactorily in the factory in August 1949. This would make it America's first working stored-program electronic computer, though not the world's first as that honor had gone to a British project (the Manchester "Baby") a year earlier. However, the BINAC project seems to have rather fizzled out after its delivery to the customer. Some Northrop engineers later claimed it "never worked" after it was delivered; others said it worked well enough for the purpose. Part of the disagreement may have been due to an ambiguous specification that referred to the ultimate requirement for an airborne computer. The former Eckert–Mauchly engineer Art Gehring believes there was never any hope of the BINAC itself being used in flight, as it was too big and would not cope with the vibration—it could only be used for ground-testing the Snark's guidance system. However, it fulfilled another useful purpose, demonstrating steps towards completing the full UNIVAC that allowed the Bureau of Standards to release more money to the company.

While the UNIVAC project was getting under way, Kay McNulty had kept in touch with John Mauchly. "I stayed down in Aberdeen but I saw him occasionally because his business was mostly with the government and to get to Washington, DC, from Philadelphia you have to go right through Aberdeen. He made a point at first of stopping to see how the ENIAC was coming along, if it was being put back together right. Then after a while he decided it was no longer as interesting as seeing me, and we decided to get married in '48. During the war his wife had died [in a swimming accident] and he had two small children. His mother had sold her house and moved in to look after the children, but he had so much to do it was very important to have someone at home and I was that person."

With her marriage the new Mrs. Mauchly returned to Philadelphia to raise more children, seven in all. But she continued to play a big part in her husband's work. "He talked to me all the time: I was more or less a sounding board. He had endless endurance and liked to stay up at night working on his ideas."

By then Betty Bartik was also back in Philadelphia, having left the ENIAC team in Aberdeen to join the Eckert–Mauchly company, and one of her tasks was doing the round of colleges to recruit programmers for the growing company. Art Gehring was one of them, an ex-serviceman who had gotten a free education through the 1944 GI Bill that subsidized college education for returning veterans, in his case a master's degree. He started on the BINAC, but it was near completion and he soon went on to UNIVAC work. "We began to think there might be a lot of applications for this that we didn't see, so we invited people from Prudential Insurance, the Census Bureau, the navy, publishing— about seven or eight different disciplines—to visit our plant and just let them see what we had and talk to the people that understood the equipment and the programming. And do you know, they came and they started to be very enthusiastic about it, and in fact Prudential actually funded part of our work they were so enthusiastic. Some of them could see that if they didn't get in on this, other people were going to get these machines and they would lose out!

"After that we were all very heartened and we began to set up procedures where we could build these machines like in a production line. It was an old mill, but we needed more space, so we rented another old building where we started to do the manufacture. Up to that point it was more by hook or by crook, but now we developed manufacturing procedures, which made it possible to really build a lot of them in a short time. Of course, this was new to all of us. It was an exciting group to work with. I feel so sorry for people today that I don't think have anywhere near that excitement and drive that we had. It was much more than a job; it really was."

Eckert and Mauchly were fortunate to have a dedicated team who didn't worry too much about unpaid overtime, or even on occasion no pay at all. But there were some irritations, says Gehring, demarcation in particular: "Engineers and programmers were quite separate to begin with. People were going to have to operate the UNIVAC that didn't know anything about what was inside of it, so Jean [Betty] Bartik and I were asked to consider what type of function they should have to be able to operate it. Well, once we started looking into it we began to say, 'We can't answer that until we know how the machine works.' So we looked at some of the diagrams and we started finding mistakes in the design. Eckert heard about this and next thing he goes to Mauchly and says, 'I'm moving those two people over into my department and they're going to start working on the logic of the machine,' so that's how we got into the engineering more than the programming."

There was no lasting resentment of the interlopers and in fact Betty Bartik remembers it as the best team she's ever worked on, even though they had no job security at all: "We didn't care. I would have gone anywhere with those guys. It was fun! It was so exciting! I've never had such a unified job environment. We knew we were pushing back the frontiers."

However, as if the financial insecurity wasn't enough of a handicap, they were hit with another quite unexpected problem. This was a couple of years before the febrile days of McCarthyism, but the hunt for Un-American Activities was already in full swing. "They had got contracts from the army, the navy, the air force, the map service . . . about six government computers, which was just great. And all of a sudden they got notice that they were all being canceled as Mauchly was a 'security risk'." Kay Mauchly still can't understand why, as "he was the least political person you could ever meet. It named four people—Mauchly, and two of the very best engineers, and Mauchly's secretary. They were working hard on the census machine at the time and Mauchly had to leave the plant and not see his own machine. The two engineers left and formed their own companies. He fought it and he had to hire lawyers, but it took about two years to get himself cleared. And there was nothing except that they said that working in the plant there were some people who were 'pinko,' meaning they had communist relatives or something. But there was no proof against Mauchly— the worst they could say about him finally was that he was eccentric."

By 1949 John Mauchly was cleared of "pinkness," the Census Bureau was still on board, some military contracts had been saved, the UNIVAC was making clear progress and Henry Straus was bankrolling the gap between income and expenditure. It must have seemed to Mauchly and Eckert that the worst of their problems were over.

Then, on October 25, 1949, Straus was killed in a plane crash. It is impossible to overstate the scale of this blow to the whole company. Betty Bartik says everyone was "crazy about Straus." With his death, the rest of the top management of the Totalisator Company just withdrew the money. "We went to the banks and they said, 'Well, what do we need computers for? Who needs computers?' Bankers are not very clever or creative and they couldn't see that these computers would ever benefit them. And the whole venture-capitalist system wasn't in place at that time. They needed someone like Goldstine." Herman Goldstine had been the crucial advocate of Eckert and Mauchly's original proposal for the

ENIAC, but of course he had split from them several years earlier and was now well established at the IAS in Princeton with John von Neumann.

Mauchly launched a desperate search for another investor. Gehring says they even went to Watson, the boss of IBM, but he didn't want to buy them. That approach must have been hard for Mauchly to swallow, and rejection even harder. Meanwhile the Tote executives were making their own moves to recover the $400,000 still owed to them. Kay Mauchly says they "knew nothing about Eckert and Mauchly. They were just very wealthy men who were interested in horse-racing. They immediately said, 'We don't want this company owing us money so we've got to get rid of it.' It turned out they had their yachts down in Palm Beach, Florida, alongside James Rand, president of Remington Rand. They made some sort of a deal between them that they would sell the 40 percent they owned to Remington Rand. Then they went to the engineers and said that Remington Rand was going to buy the company and they would buy in the engineers' 11 percent share. So they had 51 percent of the company stock, and they had Eckert and Mauchly over a barrel. They had to sell their stake for $75,000—they had no other source of money and the $400,000 had to be repaid immediately if they didn't sell up." The deal was signed on February 15, 1950, with Remington Rand taking over the whole company.

At least Eckert and Mauchly were promised financial security and they could now focus all their efforts on getting the UNIVAC into production. The company continued to grow and, on February 12, 1951, Jim McGarvey left his job at Western Union to join the Eckert–Mauchly Corporation, at the suggestion of a friend already working there: "I went for a job as a technician, but they said why not take a job as trainee operator for the UNIVAC 1? I was told it was something like a giant calculator and that interested me right away. There was no actual completed computer when I went to work there, but very shortly after that there was one ready—that was the computer for the Census Bureau. We had been working at our desks, studying some of the circuitry and the logic, and people were writing programs, but they had nothing to test them on. Time was at a premium; everyone needed time on the machine. We did work around the clock and I put in some 24-hour days. It was a very exciting time."

By the spring they had the first UNIVAC fully operational and, on March 31, 1951, it was accepted by the Census Bureau (it was formally dedicated in June). But as Kay Mauchly recalls, "The Census Bureau were afraid of moving the only UNIVAC in the world in case they had any trouble with it—what would they do? Bring it back? This was a great big room-sized machine! So they decided the best thing to do was just keep it here in Philadelphia, and keep it running here. At the same time the company could use it for testing the other machines that were being built. They had UNIVAC numbers 2 up to 7 all on the floor at the same time, but the Census Bureau UNIVAC 1 was the standard to compare against. So it was a benefit to both of them. For years the census results had been late and this time with an electronic machine they wanted to make sure the census was done on time." It was a good move to leave it where it was. The 1950 census returns were coming in and UNIVAC quickly started earning its keep. It stayed there for almost a year before the workload eased and the bureau felt they could risk moving it to their offices in Washington.

Meanwhile the UNIVAC got involved in a scheme to predict the result of the 1952 presidential election. It's quite likely that Jim Rand, with his flair for publicity and friends in high places, had something to do with it, while some people attribute the idea to another Remington Rand executive, Arthur Draper. Whoever deserves the credit, it was inspired—no computer had ever been used to predict an election result before, not even in a local election away from the spotlight. Art Gehring says, "The thought was if we can develop a program that would take a very small number of inputs and make some sort of a projection this would be a big step up as far as the publicity for the machine was concerned. So we hired a consultant, Dr. Max Woodbury, who was a Ph.D. statistician from the University of Pennsylvania, to develop the equations that we could then program to make the prediction. I guess CBS got involved—Walter Cronkite was their head honcho in the news department then and he was going to make announcements about what the machine was doing." Jim McGarvey also remembers it vividly: "I was the operator, Steve Wright was the lead programmer on that system for predicting the election, and he had a couple of people working with him."

This was actually the fifth UNIVAC off the production line and it was destined eventually for the Atomic Energy Commission. Among the many problems they had to solve was CBS's desire to have a UNIVAC in the studio. Art Gehring and colleagues found a solution: "Originally we thought the machine was going to be up there in New York, but we didn't want to take a chance on moving it. So we said we'll dummy up a phoney control panel in New York, and we'll put some circuitry in the back. It'll flash lights—that was a big thing in those days, flashing lights—and we'll keep the machine down in Philadelphia and we'll just communicate by telephone. That was the plan." The dummy console was very realistic and probably fooled most of the audience.

A bigger problem was the program itself. This wasn't like any other application they had tackled, and a presidential election was about the biggest night of the television year. "A lot of pressure got built up because this fellow who was supposed to come up with these equations kept saying he didn't have them quite yet, and we kept getting closer and closer to the election. So we were really down to the wire on that program. In fact I'm trying to think if we even had a chance to test that program out; if we did, it wasn't for very long," says Gehring, while McGarvey was right in the thick of it: "When the early returns came in we had a bank of Unitypers that typed the data onto metallic tape and we developed a spool of raw data like that. I just operated the computer but I was on television that night. The results were coming out on a teletype near the console and they came out as '00–1.' That 'zero-zero' was really a hundred but they had only allowed two places for the actual percentage so it was actually '100–1' [odds on an Eisenhower victory]. Steve Wright and the statistician from the University of Pennsylvania were not too sure of themselves, so they put a 'fudge factor' in and then it came out 10–1 instead of 100–1, because a lot of people had predicted a close race. Eventually when the result was confirmed [a landslide for Eisenhower] Art Draper went before the TV cameras and told them we had come out with the right prediction in the beginning, but we had put a fudge factor in because we couldn't believe it."

This episode brought computers in general and the UNIVAC in particular to the attention of the general public. No longer the mysterious and slightly alarming "electronic brain" of newspaper articles and the occasional black-and-white photograph, a huge audience had now seen one in operation, and seen it confound the pundits by correctly predicting their next president would be Dwight Eisenhower by a landslide when mere humans were expecting a close finish. "UNIVAC" became synonymous with "computer." Its iconic status in the 1950s was revealed in all kinds of ways and Jim McGarvey, who runs a newsletter for former employees, remembers one in particular: "A bra company wanted to use UNIVAC as the backdrop for an advertisement. There was a lot of fun that day when the model was in, actually wearing a bra, and the photographer was taking her picture in front of a UNIVAC, and of course work stopped in the whole place, everyone standing round watching this photo shoot. It was a funny situation."

The takeover by Remington Rand and the later merger with Sperry was far from happy in the long term. With the great publicity for the UNIVAC, and its potential to revolutionize so much of business, Betty Bartik reckons "Remington Rand should have sold UNIVACs like crazy but they were just so arrogant, so stupid. I went to Washington and worked as a trainer of Census Bureau programmers and after that they didn't really know what to do with me. So they used to send me out with salesmen to talk about the UNIVAC, and they would sell typewriters and accounting machines. Nobody in the Washington office but me knew a thing about UNIVAC, and this was the biggest market in the world. They used to talk about all our staff as 'those dreamers.' Well, they weren't dreamers; they were brilliant people. These salesmen had made a lot of money during the war because you could sell anything then—it didn't mean they could *sell*."

Even when Betty Bartik found the perfect application, the sales staff were too short-sighted to see it. "The Navy Supply Office had a big inventory problem. They had been using a Kardex system, but Remington Rand was about to be kicked out, so they decided to let me see if it could be a UNIVAC application—well, it was perfect. So I had it all flow-charted for the meeting with these guys, commanders and lieutenant commanders—they were the guys in the navy who really do the work. The salesmen droned on all day about their Kardex system and just gave me half an hour at the end of the day to describe the UNIVAC. Well, these people went ape, they were so excited. One would make an objection and another would answer—I was just an observer. They bought it, but Remington Rand never let me give another presentation. It was the most despised job I ever had. Our work was betrayed, by fate, by

Goldstine, by von Neumann, by a lot of people." This was the company she had praised, just a few years earlier, when she said she had never worked in such a unified environment.

Art Gehring also believes Remington Rand missed a wonderful opportunity, and with IBM hovering around they didn't get a second chance. The company "didn't have the competency and the numbers of staff. When the technology began to take off, IBM had a huge number of engineers who were skilled in electronic circuit design, whereas the original UNIVAC group hardly expanded at all. If Eckert said, 'OK, here's our next machine—it's going to consist of a drum, magnetic amplifiers, new tape, and a new printer.' That's what we built. But IBM could develop five or six different products, and if some of them sold, fine, and if they didn't, they scrapped them. Those were the big reasons why we lost out, technology and marketing." Twenty-five UNIVAC 1s were sold, a respectable total, but one that was soon overtaken by IBM. Nonetheless Gehring says, "In retrospect I'm proud I was there and participated."

John Mauchly was tied to the company for 10 years. His wife, Kay, says that he felt the people who bought it had little appreciation of what had gone into it. "Once they had gotten all these contracts, he thought he could now engage in what he most wanted to do, and that was setting up a department for developing programming languages in order to teach people how to use this machine. This was 1950, long before IBM got into it, but he could see it was the future." He already had working for him Betty Snyder, who became one of the first to devise such a language, and Grace Hopper, who had joined from Harvard. "However," says Kay Mauchly, "as soon as the Sperry Corporation took over Remington Rand [in 1955], they said, 'We'll set up a programming department in New York and any customer that wants a UNIVAC can come to us and we'll do the programming for them.' That was the last thing Eckert and Mauchly wanted—if you're planning to sell millions of computers you don't want to have to program each one for the customer. And they said to Mauchly, 'If you want to stay with the company, you'll have to move to New York.'" So John Mauchly got out of what was now Sperry Rand as soon as his 10-year "handcuffs" expired in 1960. He formed a construction management company that made good use of computerized systems and ran it for about another 10 years. It was never very profitable and, says Kay, "He never made anything that made money—too bad!"

The ABC computer came back to haunt Mauchly in the 1970s. Sperry Rand had acquired the ENIAC patents from Mauchly through Remington Rand's original takeover of the Eckert–Mauchly Corporation. Some computer manufacturers still paid royalties to Sperry Rand for the use of some of the principles of computer design derived from the ENIAC. One such company, Honeywell, didn't want to continue to pay royalties to a competitor, so they decided to challenge the patents on the basis that the ENIAC was derived from the ABC and that John Mauchly had got his ideas from John Atanasoff. Litigation started in 1967 and took four years to get to court. The hearing lasted from June 1971 to March of the following year and among the many witnesses were Mauchly, Eckert, and Atanasoff (Clifford Berry had died almost a decade earlier). It took another 18 months for Judge Earl Larson to reach his decision, but when he did so it was clear—he invalidated the ENIAC patents. Parts of the Larson judgment were pretty stark, saying that Mauchly derived from the ABC "the invention of the automatic electronic digital computer" claimed in the ENIAC patent, and much more in the same vein. Other parts of the judgment were kinder to Mauchly, saying, for example, that "the application for the ENIAC patent was filed by Eckert and Mauchly, whom I find to be the inventors," though he went on to judge that the ENIAC was a group effort with "inventive contributions" by other members of the team at the Moore School.

So is John Atanasoff the "Father of the Computer," as some assert? As we've seen, the ABC had a short life that ended in a basement and ignominious destruction. Both Atanasoff and Iowa State College abandoned it and it never solved any real problems. Probably only the connection with Mauchly saved it from oblivion. The ENIAC was not a simple derivation or even a scaled-up version of the ABC. It may well be that meeting Atanasoff first prompted Mauchly to look at digital electronic computers as a possible answer to his weather forecasting problem, even if he was too grudging to admit it. But the logic of the ENIAC was a translation of the mechanical calculators that he was familiar with, including decimal operation rather than the ABC's binary. Much of the ENIAC's conception and design was undoubtedly his own, even if not enough to save the patent. Moreover, in the wider sense of who made the greatest contribution to the emerging science and technology of computing, Mauchly has to get the verdict by a long way. The ABC was an interesting prototype that achieved little in its short life. The ENIAC and the UNIVAC solved many of the greatest problems in applied mathematics in the 1940s and 1950s, from essential ballistic firing tables, through to the design of the H-bomb, rocketry, and saving the census, to founding the election prediction industry.

In contrast, Atanasoff had another chance to contribute to the development of the first proper computers when he became head of a new computer division at the Naval Ordnance Laboratory (NOL), formed in 1945 with a brief to build a stored-program electronic computer. In the judgment of the computing historian Professor Michael Williams, this was "properly funded by the navy, it had the tacit backing of John van Neumann, and it presumably had the funds and the connections to obtain scarce material and personnel. However, it never became a serious development project and was eventually abandoned when von Neumann convinced the military that it had no prospects of becoming a reality."

The reasons why von Neumann started out as a supporter and ended up closing the NOL project down are starkly laid out in Calvin Mooers's written account of working for Atanasoff, whom he describes as charming, scientifically solid, and good at dealing with bureaucrats. But he would "elastically restate" work assignments between one project meeting and the next, and "when it came time to make the decision . . . then with all pleasantness and cordiality, Atanasoff would completely stonewall and evade making any decision or commitment whatsoever!" He had a stock of diversionary topics that became familiar to his staff, of which the strangest was the health benefits of goat's milk. He spent much of his time on his other duties as simultaneous head of the acoustic department, instead of giving that up and devoting himself to the computer project. Most bizarrely of all, he never mentioned his earlier computer to his staff. As Mooers points out, "If the Atanasoff cabalistic formula at Iowa for the design of a computer was so potent that, when the formula was 'stolen' by Mauchly, it quickly resulted in the building of the ENIAC, then why didn't the same formula work under Atanasoff's own direction in an official well-funded and well-backed project at the naval laboratory?" It's a question that Mooers answered quite simply when details of the ABC were finally revealed in the 1970s. He concluded that, even if all that information had been available to the ENIAC team in 1945, it would have been of no help to them, because it was already obsolete.

Mooers's use of the word "cabalistic" follows on from his criticism of the court process as a way of deciding who "invented" the computer. He was a witness at the court of Judge Larson and he was scathing about the Honeywell counsel's persistent question, "Who told you how to build a computer?," arguing that "the implication was that there was some mysterious critical secret about building computers, that one was helpless in building a computer without this secret, and that if you were active in building a computer, then someone must have leaked the precious secret to you. Maybe this is the way lawyers and judges think about creativity in science and technology. Certainly this mode of thought would explain some of Judge Larson's conclusions at the trial." As Michael Williams concluded, "What is true in the law often has little or no relevance to historical fact."

As well as his failure to make a go of the NOL's computer project, shortly after the war John Atanasoff also turned down an opportunity to return to Iowa State College, where he could have revived the ABC if he still believed it was important. Instead he stayed in the defense industry, and in 1952 he founded a company of his own, the Ordnance Engineering Company, coming up with a range of related and unrelated inventions, from mine detectors to farming gadgets, many of which he patented. He was undoubtedly a great inventor, but hardly the unchallenged "Father of the Computer."

On the other hand, how much of the ENIAC and UNIVAC projects would Mauchly have achieved without sight of the ABC? John Atanasoff probably got it about right when at the end of the court case he said, "There is enough credit for everyone in the invention and development of the electronic computer."

By 1972 John Mauchly had a much more serious problem. He had been diagnosed with a blood disease that was to prove fatal. He died on January 8, 1980, at the age of 72, but until then, says Kay, in spite of the illness and his years of frustration working for Sperry Rand, "He was very happy. His life was very full and he had so many interests of all kinds that he was never a sad person." His long-term collaborator Presper Eckert stayed with the company, which eventually merged with Burroughs and became the Unisys Corporation, until 1989, but he too enjoyed only a few years of retirement before dying of leukemia in 1995. Eckert left behind patents on 85 inventions, nearly all of them electronic devices of one kind or another.

CHAPTER 3
_____

SALUTING THE MOOSE

Some of the surviving engineers from the Rand 409 project reunited in 2001 under the austere gaze of the moose in the barn at Rowayton, Connecticut (author's photo).

In 1997 Erik Rambusch, a management consultant, started looking into the origins of the community center in his adopted town of Rowayton in Norwalk, Connecticut: "I was reading a local history and it implied that the UNIVAC was built here, and I knew enough about computers to know that was not correct. It was made in Philadelphia. So when I had some time I started checking with people in town." As the story began to unfold he found that an early computer had indeed been developed in Rowayton, but it was actually an almost forgotten model, the Rand 409. To recover its history, he got together as many surviving members of the original team as he was able to trace and recorded their memories.

Rowayton is a peaceful, discreetly affluent town on the shore of the Long Island Sound. It is hard to discern anything that hints at its illustrious part in the history of computing. The Rowayton Historical Society has just a few remnants of the Rand 409, including a couple of the interchangeable chassis that carried the vacuum tubes and other electronic components; these are the only known surviving relics of what were once large machines. But the Rowayton community center and library are housed in a building that was, in the late 1940s and early 1950s, the workshop in which the Rand 409 was designed and built. Earlier still it was a barn and stable, with a hayloft upstairs and room for a tractor downstairs, and storage for farm implements.

More than half a century before Erik Rambusch began his

investigations, a middle-aged engineer called Loring P. Crosman went to James Rand, the president of Remington Rand, with what might have seemed an almost impossible proposal. It was 1943 and Crosman had twenty years' experience with mechanical tabulating machines like those made by Rand's company. Although Crosman wasn't an electronics engineer, he could see the potential of electronics to revolutionize the industry and he had attended a course at Columbia University to try and come to grips with the new technology. Even so, the leap of imagination required to conceive of an electronic machine that would take in office data, perform programmed calculations on it, and print out the results was quite remarkable. This was after all at a time when the only working electronic "computers" (e.g., the Colossus code-breaking machine at Bletchley Park) were operating under a veil of secrecy in Britain, the ENIAC was just beginning construction, also in secret, for military use in Pennsylvania, and the virtually unpublicized Atanasoff Berry Computer was gathering dust in a college basement in Iowa. All these computers were designed to solve a particular problem or set of problems, but Crosman's vision was a general-purpose computer for use in commercial companies. It is hard enough to separate the competing claims for "the first computer," and almost impossible to work out who first thought up the idea, but Loring Crosman must have a good claim to be the first to conceive of a *business* computer.

Certainly he convinced Jim Rand. At the time Crosman worked for a smaller rival, but he had such a well-formed plan that Rand hired him and put him in charge of a new product development department in the Brooklyn plant, though Rand's plan was to build a new facility in Rowayton. He had his eye on the Rockledge estate, which had been built up by the steel and shipping magnate James A. Farrell and was now for sale. Rand bought it the same year, 1943, and turned the main house into his corporate headquarters.

Twenty years later it would become the Thomas School for Girls and today it is the headquarters of Graham Capital.

Rand also wanted to tear down the barn across the road and build a development center on the land, but when residents successfully objected to this, he built his Laboratory for Advanced Research a few miles away at 333 Wilson Avenue in South Norwalk. That was where the fledgling Crosman team moved in 1946.

When Crosman first came up from Brooklyn he had only a small staff—a mathematician, an electronics graduate, and some part-time technicians—but they had already made considerable progress in developing the design for what would eventually go on sale as the Rand 409. In the meantime Crosman had found time to develop some other new products for Remington Rand, including, rather less memorably, a toaster. The long search for the perfect toaster has defeated many an inventor.

Evidence for the progress Crosman had already made on his computer came when he arrived in a removal van from Brooklyn and unloaded a large piece of plywood with modular breadboards assembled on to it ("breadboards" are reusable circuit boards on which you can quickly build experimental circuits and alter them by just plugging components in and moving them around). So from his early days in Brooklyn, Crosman was thinking about a modular design based on small subassemblies. When the computer went into production, this would make servicing on the customer's premises much easier.

These breadboards formed the basis of the first prototype and some time in the following year, 1947, the second one started operating. Model 2 wasn't yet a practical computer that could be put on the market. It probably just demonstrated that the electronics would perform arithmetic reliably on data entered at one end and display the result at the other. The basic building block of "Crosman's logic," as the team referred to it, was a decimal counter, one for each of the ten digits that the computer could handle. These ten counters made up the "accumulator" and this was the guts of the machine, capable of holding numbers up to 9,999,999,999. Like the ENIAC, this was all in contrast to most other early electronic computers, which used binary counting (0s and 1s only) with decimal numbers converted to binary at the input stage and back to decimal at the output (see Appendix B for more information).

The first two models were not programmable either, but they were a solid foundation for the next step. Model 3 was very close to the production version—it was programmable and it could be used as a demonstration model for potential customers.

Now things began to move more quickly. Rand had been alarmed when his directors had accidentally walked in on another project at the South Norwalk site. He didn't want news of future products getting out prematurely, and he didn't want the same thing happening to the computer. Fortunately he still had the barn at Rowayton, so he decided to turn that into the project office while preserving the exterior to keep the locals happy. Over the winter of 1947 Crosman's team moved into the barn and promptly started expanding.

This was a good time to recruit. Millions of men had left the armed forces, many with electronic skills learned under the pressure of the war effort, others taking advantage of the GI Bill to get a free college education. So when they started recruiting for the computer project, there was no shortage of qualified and talented candidates who were also used to getting things done.

Mike Norelli was one of the half-dozen technicians who built the first machines. A graduate of the RCA Institutes, a respected technical school in New York (RCA made electronic goods like tubes and radios), Norelli was working in an age when his seniors saw no need to tell the lower grades what was happening. He says, "I built one of the first machines here and of course I didn't know what it was. In our spare time we started taking the schematic diagrams and going through them to get a better understanding of what it would do." As the group grew in size, "it got to be, oh boy, it was like a brotherhood! We would help each other in case someone got stuck on a particular activity. You always had somebody you could talk to, and there was always someone who had a bright idea of how you get around a particular problem."

Gordon Chamberlain started work at Rowayton back in 1947 and stayed with the firm for 17 years. "We had a ball! The first machine we made didn't look at all like what we had later, but we made it practically from the ground up: we made the frame for it, we made everything ourselves—it was a lot of fun. We did a lot of things that nobody else had ever done. We made what we called a 'printed circuit' that wasn't printed at all, but it was the grandfather of the printed circuit! We played around. We used to go out on the field at lunchtime and fly gliders that we made ourselves. It was a fantastic place to work. Many times it would be dark when you went home, because you'd forget the time. I'd go home at eight or nine at night and my wife used to get mad at me—you didn't call me, you didn't tell me you weren't going to be home for supper, that sort of thing—but it was because we were having such fun working here. You'd get something going and you'd just forget about everything else but working on it."

Another graduate of the RCA Institutes was Bill Wenning, who, with Erik Rambusch, has done a great deal to recover the project history. In late 1948 he went for an interview with Crosman, who offered to take him on immediately. It was just two days before Christmas, so he would get holiday pay after just two days' work, a generous gesture that he still recalls warmly more than half a century later. James Rand didn't tell him much about the project:

"He just said that we were developing a new machine for Remington Rand." It was only when Wenning started talking to the other engineers that he began to get an inkling of what it was: "I could see that there was some automation going on, but I had no idea how to classify this, as a calculator, or a computer—which wasn't a common term in those days—and so it took me a while to understand that this was different. But I had no idea then how important the machine was. In fact I believe none of us recognized that until many years later—it's only when you look back in time that you realize."

At this point the group was still small. "There were eight people in this building at that time: two engineers and people who were assembling parts for the next model. There were draftsmen and a very clever mechanical engineer who was the manager of that group. And there was Mr. Crosman himself, Loring Crosman. They were all very dedicated people and very clever, down to simple little combinations of mechanical things that you'd never seen before."

When Wenning joined the company the team was well established in the barn at Rowayton, and it was a memorable sight. "When I first walked in the door I noticed that there was a moose head up there, and I said, 'Where did that come from?' The answer was, 'Oh, he was here when we got here.' Probably the previous owner was out shooting one day and found himself one of the great attractions to this barn. Everyone kind of felt that it was part of the family. When I came back here on my second project, ten years later, I looked at it and I said, 'He's still here!' And I flipped him a little salute and got back to work." The moose head is still there now, a huge animal dominating the meeting room, usually with a cigarette in its mouth, in defiance of the adjacent "No Smoking"sign.

The place was still recognizably a barn, though Wenning says most of it "had been upgraded to a 'clean area' and freshly tiled.

Downstairs where the horses had resided wasn't as clean and that was the same floor as the computer. There were four or five stalls that you can actually still identify if you walk through and they were adjacent to the work area, but the working area was clean and up to date and very nicely kept too." Gordon Chamberlain reckons that "each guy had his own little stall, I mean horse stalls too, and I swear you could still smell the horses. Of course, there was no air-conditioning, so we had the big barn doors open all summer. It was a great place to work though." Eager new recruit Jim Marin wasn't so sure about that: "I'd always say here I am with a college degree, I've been an officer in the army and now I'm starting my civilian work and they put me in a horse's stall!" It wasn't a very secure building either but it hardly seemed to matter then. Sometimes when the team came in on Monday morning they would find signs that children had been in, playing around the computer. Many years later Mrs. Wenning's family recalled doing just that in their childhood.

This happy time also had pressure, due in large part to Rand's recruitment in 1948 of a retired four-star general, Leslie Groves. He had run the huge Manhattan Project to develop the first atom bombs and he was brought in as vice-president to oversee the rapidly expanding product development facility. There seems to have been some impatience at the pace of Crosman's work and Groves introduced a technique he had used with considerable success in the Manhattan Project. He set up a rival team in the same building, the barn.

An engineer by the name of Joe Brustman led the second team. "He had his own engineers and they painted a white line between this room and what is now the children's reading room. They didn't talk to one another!" Bill Wenning knew "their goal was to try and build a better machine using the same logical principles that our machine was based upon and do it in a form that had a better end-product—that could be manufactured less expensively, with better performance, but with the same characteristics in terms of how you operated it and what it was capable of doing. We did recognize that Loring Crosman was disappointed. You could see it in his face when the people arrived. The rest of us kind of walked around for a little bit and then went back to work and there was a change in the way we looked at things over the next few days. Most everybody got angry, not outwardly, but I think everybody felt 'We're not going to let it happen. We're going to get our job done better than they are.' And that's the way it finally turned out to be. It didn't take long for people to stop going out to lunch, staying an extra hour or two over the time they were already spending, so we're talking about maybe 10 or 11 hours a day, and coming in on Saturdays for a little bit. It happened because of Loring Crosman. I think everybody wanted to make him a success, which we did do."

So Brustman's brief was to implement Crosman's design in his own way, in the hope that the competition would spur on both teams. And from Bill Wenning's recollection of all that unpaid overtime, General Groves's calculation was correct. It was as harsh as it sounds. Les Henchcliffe, who worked on Crosman's team, recalls, "When the decision was made to go with the Crosman machine rather than the other, we went on vacation, everybody together one August. When we came back everything in that other department had been burned up, and you could see components like resistors out in the burned area. I don't know how they got rid of the metal. They must have just burned everything and carried the metal off, but they didn't want IBM or anyone to find any circuitry or anything else, physical things that would say there was a computer being developed." Joe Brustman was unceremoniously dumped within a couple of days and "he was in tears; he was out of a job."

James Rand had also recruited another general, Douglas MacArthur, hero of the Pacific war and to most of the Remington Rand engineers, many of whom had served under him. Seeing him for the first time in the flesh was something of a disappointment to Mike Norelli, because "I always thought he was a giant and then I found he was shorter than I was! And shaking hands with him was like shaking hands with a cold fish." MacArthur was sensitive about his height, which was around 5 feet 6 inches, and official photographs were carefully manipulated. Gordon Chamberlain recalls him visiting the workplace: "I never saw him without a cameraman, never. He'd take a picture of the rest of us standing up, but when MacArthur was in the picture the cameraman always used to get right down on his knees to take a picture of him, and I realized after a while that made MacArthur appear taller than he really was." If a low camera angle didn't work, Bill Wenning says, "in some cases he chose to stand on a platform and not have that as part of his picture. He was actually the same height as Jim Rand within an inch, but most of the people he circulated with were taller than him, because there were a lot of tall people, including myself and Loring Crosman. I think there was perhaps a sense that we were all too tall for him!"

Unlike Groves, MacArthur wasn't a technical man and he didn't play any part in running the computer project. John Carmichael, an engineer, remembers a visit from MacArthur: "I was running a machine on final test when he came through. I didn't let him know I was in the marine corps, in the Pacific, because of course we didn't think too much of 'dugout' [the regular army], but actually he was a very good man I guess. But he stared into the machine and he said, 'Amazing!' or something like that—he had a great voice, very deep but he didn't know what the hell he was looking at frankly. Hell no, he was a military man." Of course, that was the point, as Jim Marin puts it: "Prestige! He knew a lot of people in Washington; let's face it, he was worthwhile. He had an office across the street from us and they had to knock a wall out so he could have his own private restroom. He wasn't going to join us in a common one."

Bill Wenning thinks there was a lot more to MacArthur than just the influence he had over Washington: "He was a very unusual individual. He spent a lot of time with the operations, trying to understand which of the businesses were good for the company. Along the way he also decided that the working efficiency at the top of the structure was not to his liking. There is a story which all of us choose to believe that General MacArthur used to arrive from his home in New York City in his limousine, and he would enter the front door of the corporate headquarters and stand there to greet all his managers as they came to work. Most of the managers would arrive much later than MacArthur thought they ought to arrive, so he stayed at the front door until they all came in and over time they came to work a little earlier every day until they reached the point where he was satisfied. I think he would do that. I really do."

In 1949 work started on turning the winner of the internal contest, Crosman's Model 3, into the first production prototype, the Rand 409. No one is certain now why it got that name. Most likely the "40" came from the 40 programmable steps it could carry out, and the "9" from the number of variables it could read in from a punched card. What everyone does remember is the size of the 409 and the heat it generated. Bill Wenning says, "We're talking about a machine that would look more like three or four refrigerators, in terms of length and height, but in terms of heat you would have to say that most of the rooms they were installed in required air-conditioners. Eight kilowatts is a lot for a small room. Because of the heat, the failure rate at the beginning of our testing was such that if we found a machine that ran for more than a couple of hours at a time we thought that was a wonderful thing."

Jim Marin recently uncovered the original specification for the built-in cooling system, which was described as an "air-conditioner" but it was just "six axial flow fans mounted in the base, blowing air up through the machine." The layout was designed to optimize the cooling effect, and this was one of Gordon Chamberlain's contributions: "On both sides of the machine the tubes faced each other horizontally, and in the middle we had six big fans blowing upwards and a grille on top to let the heat out." With 8 kilowatts of heat blowing out of the computer into the room, it's no wonder that air-conditioning was needed when it was installed in a customer's premises. They didn't have that luxury in the barn: "Normally the fans would handle it. They would cool off the tubes. In a heat wave you just forget about it. You couldn't use the computer!"

Cliff Beierle believes that "in hindsight the computer is responsible for air-conditioning in the office space. Up until the introduction of the computer everyone worked in the office, they had fans running, they had the windows open and they perspired. When the 409 came out we had to have air-conditioning to keep the machine cool and everyone started saying that the machine was more important than human life!" He recalls a typical example of the Navy Medical Supply Depot, where they built an air-conditioned room for him to install the computer, and within two years the whole site was fully air-conditioned.

The program on the Rand 409 was set by making links on a "plug-board" and the working data was stored in banks of relays. Cliff Beierle says the original relay was taken from a card sorter and "it took 18 milliseconds to set and 8 milliseconds to read back. For anyone who has a PC today [around a million times faster] this has got to be mind-boggling but it did the job because at that time people weren't in a hurry; they just wanted to get it done." A brochure from the early 1950s shows two large cabinets, about 5 feet high, joined by a short bundle of cables in a metal duct. One cabinet was the Card Sensing and Punching unit, 5 feet 4 inches wide and 2 feet 8 inches deep; the other was the Electronic Computing Unit, 7 feet 2 inches by 2 feet 1 inch. An operator would set up the program on the plug-board, then load the punched cards with the input data. The computer would start processing, maybe calculating the payroll for 3,000 staff based on wage rates and time worked, and punching the results onto more cards.

There were many technical problems to solve, some of them completely unexpected. Gordon Chamberlain and his colleagues used a lot of little neon lights as logic gates in the machine (a "gate" is an electronic switch and part of the circuit that did the arithmetic). "We were fooling around with it one night. Everybody had gone home and we were still working on it because it was fun. Somebody else left the building and thought he was the last one out so he turned off all the lights. And this computer which had been working great all day started going crazy; I mean really crazy. We had a little goose-neck lamp and one of the engineers turned it on to see what the heck was going on and the computer straightened itself right out. Turn the lamp off it went crazy; turn the lamp back on it started doing what it was supposed to do. So these little neons that we were using as a gate circuit, they were photosensitive. They would fire at a certain voltage with the light on them, but without a light they fired whenever they damn well pleased! We called General Electric, who made these neons, and asked why they were photosensitive. They didn't know they were; they'd never tried using one without any lights on. That changed the whole style of the computer, because when the covers were on it was dark in there, so we had to put lights in the computer so that the neons didn't go crazy." The use of neons was one of the points of difference with the rival team led by Brustman, who designed their logic gates with diodes; that was probably a better solution technically, but Crosman's team made the neons work.

Michael Norelli worked on the development of the chassis that evolved from the breadboards that Crosman had brought from his early experiments in Brooklyn. "Each chassis was probably 18 inches long and 6 inches across, with two sets of circuits on the front and on the back side would be the tubes." These chassis also held all the neons, relays and other components that went to make up the whole computer. It was ideal for trouble-shooting because a fault could almost always be quickly isolated to one particular chassis. Swap that for a working spare and the computer would be up and running again, while the faulty chassis would be returned to the factory or the customer's own service department and repaired without holding up the data processing.

As they refined the prototype and prepared to go into production, there came an important milestone when the machine was revealed to potential customers for the first time. It's one of Jim Marin's most vivid memories: "The engineering prototype was being demonstrated to a lot of company executives in the entrance room. General Groves was heading the demonstration with Mr. Crosman operating the punch machine, and I could hear the lecture and everything—I was on the other side of the wall. He picked up a punched card, indicated the kind of information we were feeding into the computer and described what the computer was programmed to do, a payroll problem I presume. He showed them the connecting plug-cords that programmed the thing and then he held up the card that he had punched. There was a sensing unit and a punch that actually produced the results, and I heard all this applause, people clapping. We kept our fingers crossed but it all went so good they were taking orders for this thing. I hadn't documented the production yet but they were selling like hot cakes."

With the definition of the 409 production specification came a big change for the team. Numbers in the Rowayton barn had grown to 136. They had simply outgrown the place and they moved back to Norwalk in a new larger office block, one that entirely lacked the curious charm of the barn. There was no time to dwell on their new surroundings. With the production model now a reality, they had to start selling the idea to customers for whom "computer" was still a new word. Les Henchcliffe moved from the development work, which was winding down, into demonstrating it to customers. "The machine I demonstrated was in the other building, the main building of Remington Rand on Wilson Avenue in South Norwalk, and it was the former vice-president's office, mahogany paneling, everything neat and clean. Bill Wenning and I took the development machine down there on July 1, 1950, and that's when we started demonstrating. I had a flip chart that was prepared for me by Harry Mason [another engineer] and Bill Wenning, and I had a blackboard and a pointer and I gave my demonstration to customers, who were usually tabulating supervisors or vice-presidents—as many as 20 people at a time representing possibly five or six companies. And only one time in a year and a half did that machine fail to operate. I said I'd like to go on with my demonstration and fix the machine later, when somebody said, 'Let us see you fix it now!' Whereupon I analyzed the trouble, using our indicating board which was made of little neon tubes, and I could see it was in the control chassis. I had a spare chassis in the closet, pre-tested and everything, and I turned off the power, pulled the old chassis out, put the new one back in, and turned on the filaments and the bias on the tubes and the plate voltage. It took maybe four or five minutes to fix the machine; that's all. Then I took the same punched card that wouldn't go through before—I showed it to them, no answers in it. I put it in the punch section, pushed the button, and it went right through. One of the customers said, 'Let me see the card.' I handed it to him, he read what the input was and what the answer was, and I heard later we sold to every company in that room."

Quite apart from such impressive demonstrations, the size, relative simplicity, and affordability of the Rand 409 made it a practical proposition for companies. The Rand sales literature's description of it as a "punched-card electronic computer" was a clever one. Many of their potential customers were already using Remington Rand punched-card tabulating and calculating machinery. The 409 simply speeded up these machines dramatically by adding the electronic computer to process the punched cards. So for a payroll application, for example, the customer would already have a section of clerks preparing punched cards with all the employees' salary details and hours worked, and at the other end of the process there would be printing machines and clerks converting the punched-card output into pay packets. The Rand 409 replaced all the accounting clerks and mechanical calculators in the middle. This was readily understandable to customers who had never come across computers before, it could be accommodated without excessive upheaval, and it didn't risk bankrupting the company. All this for around $100,000, compared with the $1 million price tag of a UNIVAC.

The first customer was the Internal Revenue Service (IRS) in Baltimore, who took delivery of their Rand 409 in July 1951. Gordon Chamberlain was one of the men on that historic first delivery. He recalls that even getting on the truck was memorable. "The truck had a lift-gate to bring the computer on to the platform and when it got about halfway up the thing started to sway about. That was the worst thing I ever saw. I thought, 'If that falls off, all our work has gone out of the window.' But it didn't fall; he got it in the truck and strapped it down. Then we took it down to Baltimore in a moving van with air suspension, took it down the Jersey Turnpike with a police escort, before it was officially opened. We got there and found it wouldn't fit in the IRS building. It was going on the second floor, so they had to knock out a window and a wall upstairs and pick it up with a crane and swing it in. It was a monster; it really was a monster."

The IRS took the first three machines, using them to process foreign income tax. "I think one of the reasons we had a police escort was it was a Remington Rand product and Rand had a bit of pull," says Chamberlain. That "pull" had a lot to do with General MacArthur, of course. However, the IRS had more practical reasons for choosing the Rand 409, according to Bill Wenning. "When you're doing taxes for the IRS you're doing some very complex programs and you have to do them very accurately and error-free. That was one of the features of this machine, it could not make an error, and I think that was probably one of the reasons why the IRS took the first orders from us." Indeed the Rand sales manual emphasized the fact that the computer had "a means of automatically checking each arithmetic step of a calculation before proceeding to the next step." The Rand 409 did this by reversing each step and checking the result matched before carrying on. So if it added $A + B$ to get C it would then subtract B from C to check if it got A; if it did then all was well; if not it would keep trying that step until it came out correct. Finally it punched its error-free results onto cards that were loaded onto a card reader that in turn printed out the tax returns, payslips, etc.

The other main selling point for business machines was speed. Quite apart from its accuracy, the 409 could process 125 IRS returns per minute, a huge advance on a clerk using a mechanical calculator. A number of Rand 409s also went to the army and navy, who had long been using large numbers of tabulating machines for their more mundane activities like payroll processing. They quickly recognized the potential of computers for this kind of work (and not just for sophisticated nuclear and ballistic calculations). By 1952 Remington Rand was delivering one machine every month or two. These production models were labeled the Rand 409-2 and a total of about 23 were made (including the four or five prototypes). In early 1953 an upgraded version, the 409-2R, was introduced with minor changes to make programming easier and this apparently took production of the 409 series under the Remington Rand name into the hundreds.

A few months earlier though, the UNIVAC computer had come into the public eye with its impressive prediction of the presidential election result on live television. Suddenly everyone knew what a UNIVAC was and it too was now owned by Remington Rand. Within two years the Rand 409 had been replaced by two upgraded versions under the UNIVAC brand name, while "Rand" disappeared from the computer market. It was this rebranding that led to the mistaken claim, first noticed by Erik Rambusch, that the UNIVAC was developed at Rowayton.

The modifications to turn the Rand 409-2R into the new UNIVAC-branded models were mostly minor. However, there was a significant change to the working memory, with cold-cathode diodes (another kind of tube) replacing the relays that had been used on the earlier machines. This was a change that another engineer, Jacob Randmer, had been working on for a long time (they had proved rather unreliable at first). It was still a simple "accumulator" memory for storing data during calculations; the Rand 409 series never had a true "stored-program" memory. The difference between the two new models was simply that the UNIVAC 60 could use 60 columns of data from punched cards, while the UNIVAC 120 could use 120.

Before long Remington Rand was delivering up to one computer a week to business customers and in Bill Wenning's mind "there's no question that we delivered the first commercial product." Even if the first few machines off the production line are discounted, on the grounds that they were for the government, IRS and the armed forces, Rand 409s were going to genuine commercial companies from 1952–1953 onwards. They and the later UNIVAC 60s and 120s continued to sell right through the 1950s and some 1,500 were sold eventually. The Remington Rand veterans claim that this, and the fact that deliveries started in 1951, makes the Rand 409 series the world's first mass-production electronic computer. Some would say that the lack of a true stored-program capability debars it, but that consideration aside it seems a valid claim.

With sales in those quantities, servicing became a major operation. The bigger companies tended to have their own service departments but they had no experience with machines like this. Michael Norelli was one of a group of Remington Rand engineers whose jobs were in customer support. "When customers had a problem we'd try to get their service people to identify where the problem was; then you can remove the chassis and put another one in. Well, they wouldn't take the time to try and figure out where the problem was; they'd start changing components indiscriminately! I'd walk into the place and I wouldn't know whether I had any good parts at all. So I would go through and try to figure out what was wrong, and by the time I left, all these chassis were working again. They would run for maybe six weeks; then we'd go through the same process all over again. One of the things that Remington Rand was doing at that time was taking mechanical service people and trying to introduce them to electronics. That was very difficult and after a few months they started hiring electronics people and it became easier to instruct them in how you go about servicing computers. By the time I left the company the service operations were getting better."

John Carmichael found that being a "computer engineer" earned considerable respect from customers. "When you walked into an installation that was down, the manager of the place was like 'Oh my God, come on in.' You were almost like God! He might have 20 or 30 girls on key-punches waiting for something to do, and he's paying these ladies, you know. So it was just like a production line, you had to get that machine going, and we put in some awful hours, but we fixed them."

However, it was difficult if you couldn't fix the problem quickly. "I remember Mike Norelli and I were down in the Navy Yard at Brooklyn and they were trying to get a payroll out. We worked till maybe three o'clock in the morning and we were doing all kinds of tricks electronically, and changing the whole program to get it going. And we did get it going for a while; then it broke down mechanically. And the man on the installation, the customer engineer, he didn't have the part, so I was just furious. Back we went, dragging, sweating; it felt terrible, almost like you got kicked in the stomach by a horse because you wanted to make this thing go."

A much more serious blow followed on April 11, 1954, when Loring Crosman died suddenly at the age of 61. There are few personal memories of him and that's probably a reflection of the hierarchical organization that was then the norm in offices. He was, of course, always "Mr. Crosman" or "Mr. C" to his staff, but Jim Marin remembers someone who was "very nice to work for and never demanding." He could let his hair down occasionally as well. "At Christmas time Mr. Crosman would get up and perform at our parties here in the barn. We'd get together and Mr. C, who had an affliction—a large upper lip—he would put on this wig and mimic a character from Walt Disney. He had a prepared skit and he did it a number of times and he must have rehearsed it—he was good and it was funny. This is the kind of the guy he was; he was serious at times and yet when it came to relaxing he was the nicest guy." A telephone was a privilege in those days and Crosman had the only phone in the building. "When my wife wanted to contact me," recalls Marin, "she would call me and I would tell her that it was Mr. Crosman's phone and he wasn't too happy about it. I'd try to make it brief, then I'd thank him and he'd go uh-huh." Gordon

Chamberlain recalls a man who "every penny he spent he would write it down. One time I asked what he used it for. He said he hadn't the slightest idea; he never added it up!"

It's hard to say how much Loring Crosman was influenced by developments elsewhere in the infant computing industry. Jacob Randmer's impression was that "Crosman kept himself well informed about what was going on and had very strong opinions about how he should build his machine." Certainly there seems little evidence that the design of the 409 was much influenced by the computers being developed elsewhere.

Jim Marin seems to have been closer to Loring Crosman than most. "I was working at my drawing board, and he came over one day and said, 'Gee, I had a terrible experience over the weekend, a terrific pain over my heart.' Apparently he got this attack while he was driving this tractor—he loved the outdoors, he was a Quaker and he had a large farm—then it slowly went away and he recovered and he came in Monday and he was telling this story to me. I told him, 'You know, geez, you should see a doctor.' But I was in my twenties and what did I know about heart trouble? Today I would say get to the emergency room. Anyway a week later we came in on the Monday morning and heard Mr. Crosman died over the weekend of a massive heart attack. I said, 'Oh my God.' It was very sad." Crosman had at least seen his dream come to fruition and his computer installed in one of the most important of government agencies, various parts of the armed services, and numerous private companies.

With the UNIVAC 60/120 in full production the development team in Rowayton and Norwalk wound down. This was before the era of a new model every year and there seemed no pressing need to work on a successor. However, by 1957 the first transistorized computers were reaching the market and certainly by 1960 the vacuum tube-based 60/120 was obsolete.

So a new research team was set up back at the barn and Bill Wenning returned to work on the new model, to be called the UNIVAC 1004. It was a straight update of Crosman's design, now using transistors instead of tubes, with more storage and faster operations, though still at first with a plug-board for the programming. The first production model rolled out of the factory in Ilion, New York, in 1963. An updated version, the 1005, was launched three years later, finally incorporating the stored-program concept. The 1004/5 model was another reliable workhorse, relatively cheap and simple, claimed as the first "transportable" computer and robust enough for use by the US Army in Vietnam and in the Middle East. That became the most profitable punched-card computer ever made by Remington Rand, or Sperry Rand as it was by then, selling on the order of 10,000 machines. But that was the end of the "bloodline" started by Loring P. Crosman in 1943.

CHAPTER 4

WHEN BRITAIN LED THE COMPUTING WORLD

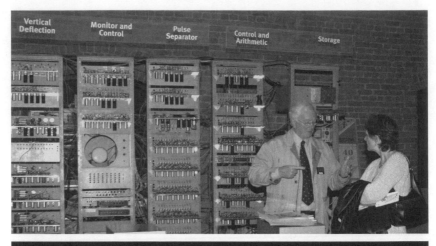

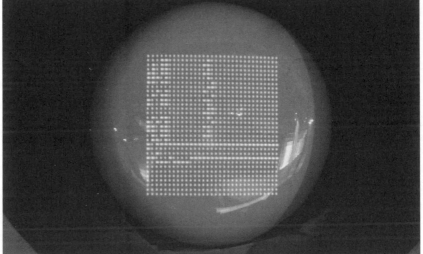

Top: Every week members of the Computer Conservation Society demonstrate the replica Manchester Baby computer in the city's Museum of Science and Industry (author's photo).

Bottom: "Immediately the spots on the display tube entered a mad dance. In early trials it was a dance of death leading to no useful result, and what was even worse, without yielding any clue as to what was wrong. But one day it stopped and there, shining brightly in the expected place, was the expected answer." This is the heart of the Manchester "Baby," the cathode-ray tube that formed the computer's memory. As well as storing the digital 1s and 0s for the computer to use, the data are visible to the operator (author's photo).

Although the United States was the first to unveil its computer to the world—the ENIAC in 1946—other countries were already working along similar lines. Foremost among these was the United Kingdom, but the British kept the secrets of their wartime computers for many years. A lot of men and women made important contributions to these early efforts but two in particular stand out, Alan Turing and Maurice Wilkes.

Turing is widely known for his "Turing test," namely "that a machine may be deemed intelligent if it can act in such a manner that a human cannot distinguish the machine from another human merely by asking questions via a mechanical link." Those with a bit more knowledge of computer history will also know him as creator of the "universal Turing machine." Some regard him as the "Father of the Computer" but it rather depends on your definition of the latter, there being as many fathers as there are definitions.

His life has been well chronicled by Andrew Hodges in *Alan Turing: The Enigma of Intelligence*. Born in London in 1912, the young Turing was a boy who behaved rather oddly: today he would be said to lack social skills. School reports were sprinkled with words like "untidy" and "inaccurate," while one of his teachers summarized him as "apt to be brilliant but unsound in some of his work." He struggled to tell the difference between right and left, drawing a mark on his left hand to remind him. His lack of physical coordination meant he did not shine at games, though he found a talent for long-distance running, something he enjoyed for the rest of his life. He struggled with driving and rarely used a car, preferring to run or cycle great distances to work, in all weather. He once told a friend that he didn't want to drive as "I might suddenly go mad and crash."

Today he might well have been diagnosed as dyslexic, perhaps even with autistic tendencies, but dyslexia wasn't widely recognized in the 1920s (though it was defined in the late nineteenth century) and autism was only described for the first time in the 1940s. However, his occasionally odd behavior didn't overshadow the mathematical brilliance that was increasingly apparent from early in his teenage years. While still in the sixth form he helped solve a problem with an illuminated cinema display case invented by the father of a schoolmate, Victor Beuttell. The problem was to evenly illuminate the whole of a poster from a single strip light at the top and the answer lay in the curvature of the glass, requiring a deep understanding of both physics and mathematics. Turing's solution impressed the Beuttells and also demonstrated a talent for mixing theory with practice that would increasingly characterize his work.

Turing won a scholarship to read mathematics at King's College, Cambridge, and in 1931 found himself in a stimulating circle of intelligent peers, being taught by lecturers with international reputations. He had a room of his own, where he could study in his own disorganized way, and his acknowledged homosexuality was less of a problem than in wider British society at the time. The following year he bought his first serious mathematical reference book, *The Mathematical Foundations of Quantum Mechanics* by John von Neumann, the mathematician who was to play such a major role in American computing. It was a demanding treatise, but Turing took to it with enthusiasm and in the summer of 1935 published his own first paper, a small improvement on an earlier piece of work by von Neumann.

The same year Turing turned his attention to another problem that was to secure his place in the history of computer science. This started with a proposition by David Hilbert, the mathematician who had dominated the field for more than three decades from the 1890s and who marked his retirement in 1928 by setting three fundamental questions for his successors. These were: "Is mathematics complete?" (i.e., could every statement be either proved or disproved?); "Is it consistent?" (so that 2 + 2 would always be 4, provided every step was valid); and "Is it decidable?" (in the sense that there is a definite method that can be applied to any mathematical assertion and eventually tell whether that assertion is true or not). The first two questions were answered in the negative by Kurt Gödel within a couple of years, but the third proved harder. That was the challenge that grabbed Turing's attention when it was presented during a series of lectures by Max Newman, another Cambridge mathematician who would go on to make a name for himself in the annals of computing.

A year later Turing published what would come to be seen as one of the most significant papers in computing history, and arguably as the foundation of computer science. It was called "Computable Numbers with an Application to the *Entscheidungsproblem*," generally referred to thereafter simply as "Computable Numbers" (*Entscheidungs* means literally "decision," and the *Entscheidungsproblem* was David Hilbert's third question). Turing's approach was to imagine a machine that could do mathematical calculations of any kind. It would do this by a series of arithmetical steps, and it would be "universal," so that the order and type of steps would define the mathematical function being performed. The data would be carried on a "tape" and the steps carried out by a controller, and he demonstrated that such a machine could replicate all the processes known to mathematics. He described a large class of what he called "computable numbers" that his universal machine could compute in a finite number of steps—and then halt. But there were other real numbers that his machine could not compute using a finite number of steps—so it would never come to a halt. He thus showed that there was no general method to determine if a computation would ever halt (i.e., if it was decidable) and concluded that the answer to Hilbert's third question was no.

This can be a hard argument to follow for non-mathematicians, but a machine that could compute mathematically is not so difficult to understand today. However, it is important to realize that, in Andrew Hodges's words, "There is not a shred of evidence that Turing ever intended to make his machine a practical reality." It was purely a theoretical construct to solve the *Entscheidungsproblem* and would probably have remained so had there not been a war.

Brilliant though Turing's solution was, he found in the course of publishing that there were two other mathematicians working on similar lines. He had to acknowledge Alonzo Church's answer to the same problem, though Turing's proof was better and he had developed it quite independently. And Emil Post had, again independently of Turing, built on Church's ideas by postulating a mindless "worker" operating on an infinite line of boxes—this was analogous to Turing's "tape," though Post did not know about the "Computable Numbers" paper. So, as Andrew Hodges concedes, "Even if Alan Turing had never been, his ideas would soon have come to light in one form or another."

The other pre-eminent name in British computing, Maurice Wilkes, arguably contributed rather more than Turing, certainly in practical terms, but is much less prominent in the popular perception. Maybe it's because Wilkes didn't have that intriguing combination of English eccentricity, public school homosexuality, mathematical brilliance, and premature death, all intertwined with the cloak-and-dagger mystique of the wartime code breakers. Although born around the same time and following a similar path through private education to Cambridge, even graduating from the mathematics faculty in the same year, there was little contact between them in the 1930s, and in later years Wilkes claimed he had no recollection of Turing before the war.

Like Turing, Maurice Wilkes was a highly intelligent mathematician, but he showed early on that he leaned towards more practical matters than the *Entscheidungsproblem*. He passed his Morse code test in the sixth form and became an enthusiastic radio "ham," and learning how radio waves bounce off the ionosphere led to a long-lasting passion for atmospheric physics. Although he took the Mathematics Tripos at Cambridge, he claims, "It was very much misnamed then as there wasn't much pure math in it. It was mostly mathematical physics." At that time Wilkes wanted to be an experimental physicist and after graduating he started a Ph.D. at the Cavendish Laboratory, researching the propagation of radio waves in the ionosphere. But it wasn't long before he, too, was drawn into the growing enthusiasm for automatic calculating machines. "It was an interesting period just before the war; there was a widespread understanding of what in principle could be done by numerical computation. The only thing was that all we had were desktop calculators—punched cards were rather little used, in Britain anyway. I did my share of computing with a desk machine, solving differential equations, and I remember when I came to write my thesis I very carefully listed in an appendix all that I had done, because I had spent many, many hours doing it and I didn't want the examiners to feel all that work was just a sentence or two, which of course is all it looked like in the end!"

During this period Wilkes was also introduced to Cambridge's first differential analyzer, which was intended as a demonstration model but "to everyone's surprise worked remarkably well." He put it to work on his upper atmosphere problems and rapidly mastered the machine, which made extensive use of Meccano and required careful attention to produce acceptable results. He soon started helping other students, including Elizabeth Monroe: "I had spent many happy hours when young building things with Meccano, this being an English boy's birthright. Elizabeth, who was American and a girl to boot, had been underprivileged in this respect," says Wilkes.

This experience with the differential analyzer led directly to a career-changing opportunity for Wilkes. "Professor John Lennard-Jones, who was interested in what we now call computational chemistry [using mathematical techniques to simulate chemical processes], took the lead and got the funds together for a full-scale differential analyzer. Lennard-Jones had great vision: he didn't just say, 'I want a differential analyzer and a room to put it in,' he said, 'I want a university department,' and so he established a complete department that only existed on paper. He was the director and I was the slave, the staff with the lowest academic rank, but it was a properly qualified university department."

This was Cambridge's "Mathematical Laboratory," although Wilkes recalls it very nearly had a rather more prophetic title as "it had been described as the 'computing laboratory' but at the last moment they changed its name, an error we had to correct a good many years later!" When it was inaugurated in 1937 he was the only full-time member of the staff, effectively setting up and running the lab under the part-time directorship of Lennard-Jones, who was already head of another department. One of the first things Wilkes did was go to Manchester to see the UK's first differential analyzer, built by Professor Douglas Hartree, who was a leading expert in numerical analysis (the mathematical discipline that was the foundation of early scientific computing). It was the start of a life-long friendship between the two men.

At around the same time, as the prospect of another major war became ever more likely, Alan Turing was developing a growing interest in the area of codes and code breaking. He devised a code that would be unbreakable "except by 100 German mathematicians working night and day for 100 years" and again he conceived of a machine that could be used to implement the code. This time, though, he was thinking of a real machine, and in 1937 he started making the first section, an electrical multiplier that used relays for performing the calculations.

It used binary arithmetic, 0s and 1s, with good reason. Binary arithmetic is at the heart of virtually all computers today, though few users need to know that, still less to understand it. Compared with the normal decimal counting numbers (0, 1, 2 . . . 9) used every day, binary numbers appear meaningless and binary arithmetic seems long-winded, yet the principle is fairly straightforward and immensely useful (see Appendix B). Binary numbers are more easily manipulated by a digital computer than decimal numbers. However, it wasn't yet clear that digital was the best method for automatic calculation by machine. The differential analyzers were analog machines, where the quantities measured varied continuously, and in this case calculations were made by the rotation of one ground-glass wheel running on another. Digital calculation uses discrete numbers that vary in steps rather than continuously, something that is more suited to electronic counting circuits.

Although Turing never finished his code generator, it was one of the first binary digital calculators and it was also part of two important personal developments: his increasing interest in turning theory into practice and his entry into the world of cryptography.

It was the war itself that brought the next big steps. The story of the breaking of the Enigma code has been told many times, but the significant fact for the history of computers is the impetus it gave to the technology in general and the particular point that Turing was now in an environment where his theoretical mathematics had real consequences. The prompt solution of a mathematical puzzle, an intercepted coded message, could and did save lives in the shipping lanes of the Atlantic. Turing also came into contact with more machines that did computations, first the "Bombe" and then the "Colossus."

The Bombe was a much improved version of the "Bomba" invented by the Poles, whose code-breaking work on the German Enigma led the way for the British (the Polish contribution has sometimes been overlooked in popular coverage—it wasn't all done by the British). The Bombe was an ingenious machine that used thousands of rotating wheels to cycle through possible ciphers that would decode the message being tested. When a potential match was made, a complex electrical circuit was completed and the wheels stopped. At that point the operators would check the result and see if it would indeed decipher other messages for the same day. False matches were common, but it was successful often enough for the timely use of decoded messages. However, it did no calculations, and it was dedicated to this one purpose. Turing played a major role in developing the British Bombe, along with a host of other code-breaking methods, and it must have encouraged his interest in calculating machines.

The first real computer at Bletchley Park, and the first large-scale electronic computer in the world, was the Colossus. The Colossus was aimed not, as is sometimes suggested, at the Enigma machines but rather at the German Army code produced on Lorenz machines and labeled "Fish" by the British. It was Turing's Cambridge lecturer Max Newman (the person who had first alerted him to the *Entscheidungsproblem*) who developed the specification for the Colossus and the brilliant Post Office engineer Tommy Flowers who designed it, its construction taking just 10 months in 1943. Its 1,500 electronic valves could each switch in around a millionth of a second, rather than the thousandths of a second required by relays. This enabled Lorenz-coded messages to be cracked in time to provide operationally useful intelligence. However, the Colossus had no memory of the kind that could store both data and program. Programming was done by a combination of switches and cabling (just as with the ENIAC) and the output appeared on a teleprinter. Six months later came Colossus Mark II, five times faster and with the valve count up to 2,400. More significantly the Mark II had an important new feature, conditional branching (just as Charles Babbage's nineteenth-century design for an analytical engine had done, though there's no evidence to suggest the designer was aware of Babbage).

While Turing is sometimes credited with the invention or development of the Colossus, he actually had no direct involvement. While nominally in charge of "Hut 8," which housed the team responsible for cracking the Enigma-coded messages, his poor administrative and personnel skills meant that his nominal deputy, Hugh Alexander, had became the de facto manager. By 1942 Turing was increasingly following an interest in speech analysis and scrambling. That year he went to the US to improve the liaison between the British and American cryptanalysis programs. But more significantly from the point of view of his developing interest in computing machines, he also spent two months at Bell Labs, where he learned about electronic speech encipherment. There he met Claude Shannon, who had followed a similar independent path towards the creation of logical machines, working on differential analyzers and writing a seminal paper back in 1937 that first described switching relay circuits in terms of boolean algebra (coincidentally the same year Turing was building his electrical multiplier embodying precisely that principle). They discussed machines that would compute, or even "think," like a human brain. For all that computer engineers would later disparage the mass media's use of the label "electronic brain," it's likely that people such as Turing and Shannon started it.

Returning to England, Turing busied himself with his own unbreakable speech encipherment project, the "Delilah," based not at Bletchley Park but ten miles away at Hanslope. It wasn't completed before the end of the war, but even at the start of that project he openly stated his intention of turning his theoretical "universal machine" into a real "electronic brain." And it was his involvement in wartime electronic machines that had brought about this advance in his thinking. Very importantly, though, this electronic brain would not be a special-purpose Bombe, Colossus, or Delilah, all of which needed re-engineering to tackle each new problem. Instead his universal machine would need only different sets of instructions to tackle a wide range of scientific problems, and the end of the war brought the opportunity to turn his idea into reality.

In 1945 the National Physical Laboratory (NPL) established a new mathematical division in Teddington and appointed as its head John Womersley. He knew of Turing's "Computable Numbers" paper, he had seen Howard Aiken's Harvard Mark 1 in operation (he called it "Turing in hardware') and he was the first non-American to see the "Draft Report on the EDVAC." He wanted Britain to compete with the Americans to build a digital computer and he recruited Turing to make it possible. Turing's report "Proposed Electronic Calculator," written in 1945, sets out in considerable detail what was soon dubbed the "Automatic Computing Engine," or ACE. In the report he describes the importance of a memory for data and program storage, and how this would be realized with acoustic delay-lines. Turing suggests they could be filled with either mercury or water, and although he was not the first to propose the use of delay-line storage (that was probably Presper Eckert in America) it was typical of him that he went into the basic physics in detail before adopting this as the preferred solution. He didn't just take as read the prevailing preference for mercury; instead he looked at a range of other fluids, including mixtures of alcohol and water.

It's also noticeable that nowhere in Turing's report on the ACE does he refer to his own paper "Computable Numbers," instead saying near the start that "it is recommended that it be read in conjunction with J. von Neumann's 'Report on the EDVAC.'" It is understandable that Turing followed good academic practice in acknowledging prior publications on the subject, but it did create the impression among some readers, both at the time and later, that his report (and hence the ACE) was derived from the EDVAC design. In fact, starting with his theoretical Turing machine as described in "Computable Numbers," and looking at his early code-breaking binary electrical multiplier, his wartime involvement in electronics, and his earlier discussions about making an electronic brain, he had the basis for the ACE design with or without the EDVAC report.

The NPL's mathematics division was not a very adventurous organization under John Womersley. Rivalry with the Post Office Research Station at Dollis Hill in northwest London (which had built Colossus and was supposed to build the ACE), combined with Turing's odd personality and the sudden loss of wartime pressure and resources, meant progress was slow. Nothing of significance had been built by mid-1947, when the NPL director, Sir Charles Darwin, canceled the contract with the Post Office and decided to build a smaller "pilot" version of the ACE in-house. That autumn, Turing, who had never really fit in at the NPL, went back to his fellowship in Cambridge for a sabbatical from the ACE.

There Maurice Wilkes was hard at work on his own design for an electronic computer. Wilkes, too, had gone through many formative experiences in the war, though none had involved either the code-breaking or primitive computing machinery that Turing had encountered. The first thing that happened to Wilkes was that his precious ham radio equipment, like all privately owned transmitters, was impounded by the Post Office as soon as war was declared. He then became one of many scientists recruited by the Air Ministry in the summer of 1939 to help develop and run the new radar installations that proved so vital to the country's defenses during the Battle of Britain and afterwards. This immersion in the technology of high-frequency pulses would prove valuable when he began to design his own computer circuits, although he says that "the most important thing was I learned how to get things done! You see one of the features of the times was 'crash programs,' particularly for the air force. When we developed some new model of airborne radar, then the air staff would say, 'Can we equip a couple of squadrons as soon as possible?' I mean it was going to be manufactured at very high speed anyway, but the crash program was *in addition* to that—you just did it right away! So I learned how to do that and it was very valuable."

Wilkes joined the Telecommunications Research Establishment (TRE) at Great Malvern in Worcestershire and worked for a time on the Oboe guidance system that allowed Bomber Harris's RAF planes to find their German targets in darkness. He says he had "certain unresolved ethical problems in regard to the bombing of industrial and civilian targets" and was glad to leave that project for work that was more clearly related to military objectives.

Immediately after the war in Europe ended, Wilkes went on a six-week scientific fact-finding tour of the American-controlled zone in Germany, which he enjoyed immensely. "I'd been cooped up in Britain all the war you see and I always like to put it this way—that a grateful government gave me six weeks holiday in Bavaria and Austria and it was a lovely summer! I got there and I learned that you could volunteer for this intelligence work interrogating German scientists. It turned out there were files of these intelligence targets and we could choose one and go off and do it. I discovered that radar was fairly well covered by other people, but my old pre-war field of ionosphere research was not at all well covered and a lot had gone on during the war, so I was able to renew my interest in ionosphere work and meet some of the people whose names I knew. They knew my name from pre-war publications or even wartime publications." It was also in effect his introduction to the United States as "I was in the American zone and it was really like being there. I always say that was my first trip to America, and I am very much an admirer of things American."

On his return to Cambridge he found that Professor Lennard-Jones, the part-time director of the Mathematical Laboratory, was tired of doing two jobs and so the university made it into a full-time role and offered it to Wilkes. His first task was to evict the military men who had commandeered the lab for war purposes and this took until January 1946. By then Wilkes had heard quite a lot about the American developments from his old friend Douglas Hartree, who had put his differential analyzer to military use during the war and thus got to know about many of the other developments in computing on both sides of the Atlantic. Hartree had toured the United States while Wilkes was in Germany, returning with news of the ENIAC and the Harvard Mark 1 in particular. In February 1946 Wilkes wrote in an internal paper that "Cambridge should take its part in trying to catch up some of the lead the Americans have in this subject" and set out his plans to get involved. That prompted Hartree to visit him, concerned at the modesty of Wilkes's proposals and emphasizing the great scale of the American projects. However, there was no way that Britain's shattered economy could match the resources available on the other side of the Atlantic. Wilkes and his contemporaries could only hope that British ingenuity would keep them in the race. The same year Douglas Hartree joined Maurice Wilkes at Cambridge, becoming Professor of Mathematical Physics, and strongly supporting Wilkes's efforts to build a computer for the university.

Events moved quickly in 1946. In May, Wilkes saw von Neumann's "Draft Report on the EDVAC" and not long afterwards was excited to get a telegram inviting him to the Moore School computer course in July and August. His frustration at being unable to get a passage to America until the beginning of the latter month is easily imagined, post-war shortages affecting shipping as much as every other area of British life. He arrived with only two weeks of the course left, so it was a relief to find that he had missed little as the first month had been devoted to basic mathematics that he was already familiar with. The final fortnight was invaluable, as along with the whole class he pored over ENIAC circuits and discussed various ways of implementing von Neumann architecture. What he didn't get was a set of plans for the EDVAC, as no such thing existed then, but he left with the general principles of a stored-program binary electronic computer clear in his mind, along with an enduring respect for Presper Eckert and John Mauchly: "They remain my idols. Eckerts of course was a brilliant engineer and he and I took to each other very much. I admired him and of course the ENIAC was a 'crash program,' he said, 'we built the ENIAC in something of a hurry,' and I knew exactly what he meant. Eckert was a great talker; he did most of the talking if you weren't careful!"

After the course, which he calls a "wonderful, wonderful piece of generosity," he spent more time with John Mauchly and he went to Harvard to see Howard Aiken's Mark 1. He was skeptical about Aiken's belief that a major role for a digital computer would be computing mathematical tables (the same purpose for which Babbage had designed his Difference Engine, over a century earlier) and Aiken's concern at the difficulty of printing this abundance of new tables. Wilkes saw instead that electronic computing would greatly reduce the need for such reference tables, but Aiken was an overbearing personality and Wilkes didn't challenge him. He did find the courage to disagree over another issue. "He couldn't stand the binary system and one day while I was there he was holding forth about this and I took him on and we went at it hammer-and-tongs and he was delighted! I think he was rather puzzled by the fact that people wouldn't argue with him, but anyway he invited me to his home, his wife invited me to dinner, he fed me highballs, and there were some other guests and we had a wonderful night. He could be very charming indeed." Still he was in Wilkes's judgment "an example of a great pioneer becoming very reactionary and all the people who were building the early computers didn't listen to Aiken."

Wilkes wasted no time in applying what he'd learned at the Moore School. "I came back on the *Queen Mary* and on the way I began to sketch out something that I might do. I gradually came to realize that, in spite of the enormous scale of the American activities, I could get started with the resources that were available to me, and I did so." This was the approach he had learned during the war: getting things done. "You have to remember that this was a country that had been at war for six years, and we were all heartily sick of that kind of thing. Everybody realized that we had to get civilian values and activities re-established, and in the university, if you had some ideas, then you got support." By the time he returned to Cambridge he had the outline of his computer clearly sketched out and he could get straight to work; he "didn't have to ask anybody, put in any proposal, arrange any budget. I was in charge and I could go ahead. The times were extremely abnormal."

During the Moore School course he had concluded that "the basic principles of the stored-program computer were easily grasped, but how they were to be implemented was, in the summer of 1946, an entirely virgin field." Much of the circuitry required could be developed from past experience because designing an electronic computer to do arithmetic wasn't regarded as particularly difficult by then, but making a high-speed memory was. This was a completely new requirement and it was the major problem standing in the way of the modern computer. Presper

Eckert had come up with the idea of using delay-lines, specifically mercury tanks, and this was readily understood by Wilkes from his radar work.

While at the Moore School he had also seen the Selectron, which looked so promising but was to hold back some of the best American projects for years. Fortunately for Wilkes he preferred the mercury delay-line option and the importance he attached to that choice is shown by the name he gave to his computer, the "Electronic Delay-Storage Automatic Calculator," or EDSAC. He was fortunate to find in the Cavendish Laboratory a man who had spent much of the war working with mercury delay-lines, Tommy Gold. Delay-lines had been vital components in radar—by delaying the radar signal from one sweep of the screen and subtracting it electronically from the next sweep, it was possible to remove most of the reflections from fixed objects and leave only the moving ones (like aircraft and U-boats). Gold was just finishing his Ph.D. and readily transferred to the Mathematical Laboratory, where Wilkes's first priority was to turn a mercury tank into a usable memory.

The principle of the delay-line could be used to create a computer memory, as a series of pulses of sound in a bath of mercury. An analogy might help to make the principle clear. Imagine standing next to a motor-racing circuit and watching the cars passing. On every lap they come off the track and through the pit-lane, where you are standing. There you can switch the headlights on or off, and you use this ability to make the cars carry a code for you, where "headlights on" equals "1" and "headlights off" equals "0." You might decide that the first ten cars would all have their headlights on and this sequence of 1111111111 would be the code to show the start of the data. Then you switch the headlights of the following cars on or off in sequence according to your program (for the purpose of this analogy, the drivers have no control over the headlights). Do the same with your data and you have all the information you need stored in a "memory" made up of car headlights circulating round a track. Read the first program instruction and data from the headlight code, start your calculations, and whenever you're ready for the next step wait for the cars to come round again and decode the next instruction.

A mercury delay-line works in a similar way except that the 1s and 0s are represented by pulses of sound, created at one end of the bath by an electronic speaker and detected at the other end by a microphone. The data pulses are sent back round a loop to the start, and can be read or changed on the way. Because sound travels much more slowly in mercury (about 500 miles per hour) than electricity travels in metal wires (about 186,000 miles per second), the computer has plenty of time to do its calculations while the pulses circulate endlessly in the mercury bath.

If the principle is hard to grasp, it was harder still to make it work in practice, so Tommy Gold's experience was invaluable. The delay-lines had to be physically large to store enough pulses and yet they had to be very precisely engineered. The EDSAC used 5-foot-long delay-lines, machined to an accuracy of a thousandth of an inch. Another problem was that the quality of the pulses degraded as they traveled round the circuit. To return to the race-track analogy, it was as if the headlights were powered by small batteries that ran down quickly and had to be recharged each time the cars passed through the pits. In a similar way the pulses in the mercury delay-lines had to be regenerated each time they passed round the circuit, and this was just one of a number of technical problems that Tommy Gold had to solve.

To the delight of Wilkes and Gold they had a working prototype by February 1947, just six months after the Moore School lectures, and this event was duly celebrated with several pints at their local pub, the Bun Shop. They marked each milestone in the same way, and celebratory visits to the Bun Shop recurred with reassuring frequency. Circuits were designed as they went along and with considerable speed; Wilkes reckons that if they'd given the same attention to every detail of the design as had been given to radar and TV sets, the project would have taken twenty years instead of two.

With Alan Turing back in Cambridge by late 1947, it might be thought that he would have contributed to Wilkes's work, but the two men did not see eye to eye. Turing's conception was of a machine with minimal hardware, relying on clever and complicated programming to achieve its tasks. Wilkes both inherited and supported the Americans' greater reliance on hardware, and wanted programming that would be more readily achievable by students and staff in general. He said later, "The machine was to be simple with no frills, except that it was to be comfortable to use. I didn't want it to be the kind of machine where the user had to know about the pulses inside it, or timings, and so on. There was to be no attempt to fully exploit the technology. Provided it would run and do programs that was enough." There were many other points of difference and they could be scathing about each other. Wilkes remembers attending the first of a series of lectures by Turing at the end of 1946 about the principles of the ACE. Wilkes says he "did not believe that computers would develop along the lines that Turing was advocating and for this reason I stopped going to his lectures." The decision to stop attending was no doubt reinforced when Wilkes dared to venture criticism of Turing's views on the design of mercury delay-line memories; Turing's response was prickly despite his "lack of competence" in Wilkes's eyes. Nonetheless Wilkes still has great admiration for all that Turing contributed to computer science.

While Wilkes's team pressed ahead in Cambridge (and the NPL dithered), others too were putting their wartime experience to good use. Max Newman had moved from Bletchley Park to Manchester

University in 1945, intending to build a computer there. He got a grant from the Royal Society, in spite of strong opposition from Sir Charles Darwin, who wanted his NPL to have a monopoly and the ACE to be *the* British computer. Darwin was even more annoyed when Manchester recruited Freddie Williams from TRE, where he had started experimenting with memory storage using a cathode-ray tube (or CRT, the display screen in a radar or television set) as an alternative to the acoustic delay-line. Alan Turing was also recruited to Manchester, in 1948, breaking a commitment to return to the NPL after his sabbatical year in Cambridge. Darwin's dream of preserving wartime cooperation and all pulling together to build the ACE was well and truly shattered, but he had only his own organization to blame. Had NPL forged ahead with the ACE as soon as the war ended, it could have been up and running by 1947, leaving the Manchester Baby and Cambridge EDSAC as mere footnotes.

Instead Manchester and Cambridge led the way. In Manchester Freddie Williams and a colleague, Tom Kilburn, were tackling the particular problem of the day, the development of a practical memory. Both the ACE and the EDSAC teams had stuck with the proven technology of mercury delay-lines, even though they were bulky, expensive, awkward to set up and relatively slow in use. The Manchester team preferred the idea that Williams had brought with him from TRE of using a CRT to store data. This would be faster than a delay-line, as the beam could be directed straight to the part of the screen where the data was stored (instead of waiting for the data to circulate round the delay-line), but there were many practical problems to solve. CRTs were not built for this purpose and they had manufacturing imperfections that would not matter at all in their intended use in TV sets but caused data errors when used for computer storage. They were also very prone to interference and an unsuppressed motorbike roaring past the building would send the data haywire. So they decided they would first build a small prototype to test the practicality of using a CRT. Formally called the "Small-Scale Experimental Machine," it was much better known as the Manchester Baby. It wasn't an instant success, as revealed in Tom Kilburn's much-quoted memory: "When first built, a program was laboriously inserted and the start switch pressed. Immediately the spots on the display tube entered a mad dance. In early trials it was a dance of death leading to no useful result and, what was even worse, without yielding any clue as to what was wrong. But one day it stopped and there, shining brightly in the expected place, was the expected answer."

That first successful run was on June 21, 1948, making the Baby almost by accident the first "stored-program electronic digital computer" in the world, although even the design team didn't regard it as a full-fledged computer. It was a cut-down design capable only of subtraction and intended purely to prove the CRT storage idea, but within those limits it was the first working electronic computer to have both data and program stored in a common memory.

The Manchester Baby's triumphant debut rather overshadowed Maurice Wilkes's unveiling of his EDSAC in Cambridge a couple of weeks later, and it was May 6 of the following year before the EDSAC ran its first full program. That was an important day because this was certainly "the first complete and fully operational regular electronic digital stored-program computer," as Cambridge University now bills it. The moment was recorded in the machine log with typical restraint simply as:

> 1949, May 6th. Machine in operation for first time. Printed table of squares (0–99), time for programme 2 mins 35 secs. Four tanks of battery 1 in operation.

Note incidentally the spelling of "programme"—the abbreviated spelling came later. Within a year the EDSAC was providing a programming service to the whole university, which it continued to do until its eventual close-down in 1958.

Thus, in 1948, Britain was at the frontier of electronic computing. In Manchester the world's first stored-program electronic digital computer, the Baby, had run its first program, and it was the basis of a more powerful model that would be one of the first in the world to go on commercial sale. In Cambridge the world's first full-scale stored-program electronic digital computer had just been unveiled; it would shortly run its first successful program and go on to give many years' service as an everyday working machine. It's worth emphasizing that this was Wilkes's design, not a copy of the American EDVAC: "It resembled the EDVAC in that it was a serial machine using mercury tanks, but that was all. Everything else was entirely different. When I was at the Moore School the EDVAC design did not exist, except maybe in Eckert's head." Also in 1948 a small team at J. Lyons & Co. was starting work on the business version of the Cambridge EDSAC, and it in turn would within three years run the world's first real day-to-day business application. Even the Pilot ACE, delayed as it had been, was about to get a proper boss who would give the project the boost it needed. The British teams were using mercury delay-lines or cathode-ray tubes for working memory and both approaches were succeeding.

In contrast the American projects had run into various problems. The ENIAC had gone into full service too late to do the urgent job it had been built for, though it did useful work for another 10 years. Its lack of true stored-program capability and its use of decimal instead of binary arithmetic limited its usefulness and meant that technologically it was in several key respects a dead end. The EDVAC was making slow progress from brilliant idea to actual hardware in the depleted Moore School, while von Neumann's own project, the IAS computer, was similarly hampered at the Institute for Advanced Study in Princeton. Several American teams were relying on the Selectron or iconoscope for storage, persevering for too many years before in most cases giving up these fundamentally flawed approaches. Eckert and Mauchly had a better technical approach with the UNIVAC, but were hamstrung by poor business skills and lack of money. Their BINAC became the first American stored-program electronic computer in August 1949 but it was another dead end, whose only real value was in generating income and experience for the UNIVAC. The Rand 409 was making good progress but it had no stored-program capability in the true sense, although it was cleverly conceived as a low-cost business machine and intended for mass production. The American machines would only start going into service from 1951 onwards.

It is hard to conclude other than that Britain's computing effort led the world in 1948, despite the crippling privations of the early post-war years. But the industrial might of the former colony across the Atlantic could not be held back for long, particularly as the British effort was directed towards single machines or small production runs whereas the Americans were capable of much more, or soon would be. Maurice Wilkes recognizes now that it was probably only ever a temporary advantage because "our intention was to make a machine that was useful, not the best possible machine, not exploiting the technology to the full, and there were specific decisions relating to that point. And we cut corners so we got a machine of some sort. Now that wouldn't have suited Eckert at all because he and Mauchly wanted to make a machine they could sell."

The British continued to make good progress for several more years. Wilkes and his team were quick to start programming. "I fortunately had experience [with] computing—running the Mathematical Laboratory before the war, where a lot of computing went on—and very few people in my position did. And so it was a feature of Cambridge that when we got the machine going we immediately started using it; there was no question of a team of engineers handing over to an entirely separate team of programmers or mathematicians. This was the second part of the game, learning how to program, and so we did some of the pioneering work in programming methodology and the term came into use later on. But it didn't really appeal much to the other groups. Also my experience made me realize that there were lots and lots of people like I was before the war, when I was a student, who didn't want to do vast big problems like weather forecasting and so on, but who did have that wretched integral that had to be done, that would have meant weeks at a desk machine. These things were well within the power of the crudest sort of programmable computer. So the idea was to make a machine that would do that."

Around the same time Francis Colebrook arrived at the NPL to take over the pilot ACE project. He was an excellent manager, who moved the mathematicians into the same building as the engineers, got them all working together, set up something like an assembly line, and got the project moving at the speed its designer, Alan Turing, had expected three years earlier.

But Turing was by now part of the Manchester team, still basking in the success of the Baby and pushing ahead with the full Manchester Mark 1, or MADM (Manchester Automatic Digital Machine), which was to be the basis for a commercial product, the Ferranti Mark 1. Turing's role was nominally head of department, but it wasn't something that suited a man with such odd social skills and when he got a gifted deputy he was glad to pass over the administration to him (just as he had done years earlier at Bletchley). In any case, according to Wilkes, "Freddie Williams took good care that Turing did not meddle in the engineering design of the machine that he and Tom Kilburn were building!'

Instead Turing concentrated on producing his programmer's handbook, a remarkable document that showed how the full-scale MADM would be programmed, with many examples. However, it was also remarkable for its lack of concessions to users. Anyone without Turing's mathematical brain would struggle to follow it, and he was notoriously unsympathetic to anyone not on his wavelength. Frank Sumner, later a distinguished professor of computer science at Manchester, was then one of the Ph.D. students who had to use Turing's programming manual. He found that, while it was full of examples, most of them didn't work. "Turing was like that: if he wrote $1/k$ instead of $1/t$, he knew it was $1/t$, so why bother to do all that proofreading?" The handbook was given a major rewrite by Turing's successor as head of software development, Tony Brooker, who added the introductory note: "Much material has been taken over and altered, or only slightly modified, from the first edition written by A. M. Turing." Frank Sumner's comment on this is that "it is the most glorious piece of understatement you could imagine . . . 'slightly modified' meant about one character in every 20." Brooker also showed more indulgence than Turing towards the novice programmer, with "a set of footnotes . . . making comments on some less obvious points in the description and design."

Turing could nonetheless be generous to those he regarded as intelligent, like Christopher Strachey. When the young Strachey wrote his first program, the longest one the Manchester team had ever seen, no one thought there was the slightest chance he could make it run. They left him to work on it overnight (a common practice when computer access time was so precious) and he astonished them the next morning by demonstrating a working program that did its intended task, then finished with a flourish by playing the national anthem on the hooter that was normally only used to attract the operator's attention. Computer-generated music was an early interest across several continents and at the end of that year, 1951, the BBC broadcast the machine playing Christmas carols.

What Turing was increasingly absorbed in was the relation between electronic computers and human intelligence. While Wilkes, like many of the pioneers, despised the media label "electronic brain" and was annoyed by Lord Louis Mountbatten's popularization of the term in a famous speech in 1946, Turing had been talking in those terms for years. Not that he thought the first modern computers had much in common with the human brain, but he was trying to visualize the extent to which computer-generated artificial intelligence might come to resemble human intelligence in the future. This was deep stuff. Philosophers had long argued over the extent to which humans have free will and the nature of consciousness, without (as is the wont of philosophers) ever deciding either matter. Some religious commentators railed against the idea of computers that could "think" (and similar concerns were holding back Soviet pioneers, rebuked for using terms like "memory" and "logic" and constrained in their development of cybernetics).

Mainstream psychology (in particular in America) at that time was strongly influenced by behaviorist models, by which it was believed that human behavior was dictated by past learning. The purest behaviorists had little regard for conscious thought, still less the subconscious (Freud was anathema to them), with only observable and measurable behavior regarded as a proper subject of study. So it is intriguing that Turing was moving towards a definition of intelligence as "the ability to learn" and experimenting with computer programs that appeared to demonstrate the machine "learning." He set that out in 1950 in a major paper entitled "Calculating Machinery and Intelligence," described generously by Wilkes as "witty and illuminating."

Unfortunately dark clouds were gathering over Turing.

Although he had come to terms with his own homosexuality, society had not. When he was burgled by an acquaintance of a young man he had picked up on a Manchester street, he innocently reported the burglary to the police. When the burglar told what he knew of Turing's relationship with the young man, the police arrested them too. Turing was charged, prosecuted, and ordered to undergo female hormone treatment; at the time many psychiatrists believed in chemical cures for the "aberration" of homosexuality. Before long behaviorists would turn to electric shock aversion therapy to "un-learn" the deviance, but Turing was spared that. Indeed he seemed to endure the whole ordeal reasonably well. However, growing Cold War paranoia had emphasized the security services' belief that homosexuals were vulnerable to blackmail. Alan Turing found himself excluded from classified work, and in 1954 he was discovered dead in his bed with a half-eaten apple on the floor. It was soaked in cyanide and there wasn't much doubt that he had committed suicide.

Maurice Wilkes had a much longer and happier career, during which he contributed much to computer science, both hardware and software. It is striking how often his name comes up in conversation with other pioneers in America, Russia, and the Ukraine, even if he never achieved the popular recognition of Turing. He was knighted in 2002 for his services to computing and when interviewed for this book in 2004, at the age of 91, he was still working regularly in his office in the Computing Laboratory at Cambridge University. Thus his working life spanned the whole development of computing in the twentieth century, from the desktop calculator to the wireless PDA. His first computer, the EDSAC, was also the basis for the world's first business computer.

CHAPTER 5

LEO THE LYONS COMPUTER

David Caminer was manager of the LEO applications team, devised
the novel teashops distribution program, among others, and still
contributes to the development of business computing in his eighties
(author's photo).

At the turn of the nineteenth century the British company Salmon & Gluckstein was the largest tobacco retailer in the world, based primarily in London, where it had opened its first shop less than 30 years earlier. During that period of rapid growth, some of the younger members of the founding families grew unhappy that their entrepreneurial ambitions were confined to tobacco products and began to look for other opportunities, but the move into catering was more serendipitous than planned.

The second half of the nineteenth century was the age of exhibitions. While these were hugely popular events, the catering at exhibitions was notoriously awful. Montague Gluckstein, son of one of the two founders of Salmon & Gluckstein, attended such shows regularly to market the firm's tobacco products and he decided they could do much better. The other members of the two families were appalled at the idea of their respected firm associating with the despised business of exhibition catering. But Montague won them over on the condition that the catering business was operated under a new name, so a relative, Joseph Lyons, was brought in to front the company.

J. Lyons and Co. was formed in 1887 and was an immediate success at the Newcastle Jubilee Exhibition. Instead of high prices and low quality, Lyons offered freshly brewed tea and coffee, cakes and biscuits straight from the oven, along with innovative sideshows such as demonstrations of cigar rolling and the world's first realistic fairground shooting gallery. The operation was so successful that the tea pavilion outlasted the exhibition.

The new company's exhibition catering business expanded rapidly and by 1891 Lyons had teamed up with another businessman, Harold Hartley, to stage an extraordinary spectacle called "Venice in London" in the huge 4-acre complex of Olympia. This elaborate re-creation included extensive waterways, 50 gondolas imported from Venice complete with Italian gondoliers, and, of course, many catering outlets. It ran for 13 months from Boxing Day 1891 until January 1893. In those first few years Joe Lyons had learned that his best chance of guaranteeing quality was to do as much as possible himself and this became one of the enduring characteristics of the Lyons company.

Montague Gluckstein looked at high-street tea and coffee shops and decided they were ripe for the same sort of revolution. At that time, as the historian Peter Bird explains, "the typical coffee shop was the old 'slap-bang' where you stood in the sawdust at a counter, put your penny down, got a cracked cup of filthy liquid slammed down in front of you, drank it and went on your way." Bird worked for Lyons from 1948 to 1976, and subsequently wrote *The First Food Empire*, the definitive history of the company.

Gluckstein believed there would be demand for a much more civilized place to sit down and be served tea and coffee, combining lower prices with guaranteed quality. The first Lyons teashop opened in 1894 at 213 Piccadilly (it is still a café, now Ponti's). Although largely modernized, it's still possible to see remnants of the original decoration, the stucco ceiling in particular.

The venture was an immediate success, from the day the first teashop opened its doors to queues that stretched down the street. By 1900 there were over 50 Lyons teashops in London and other major cities, with nine in Oxford Street alone at one time. Crucial to their popularity were the elegant and efficient waitresses, one for every eight customers (which became a Lyons standard) and recruited on appearance as well as ability. An early requirement specified a maximum waist size of 17 inches and Lyons provided tailored uniforms of long dark dresses with white aprons and accessories. Before long the company had its own dressmaking department and laundry, demonstrating its determination to control quality by doing as much as possible in-house.

In 1894 Lyons acquired Cadby Hall, once a great piano factory and showroom and situated next to their original headquarters at Olympia on Hammersmith Road in London. Cadby Hall had plenty of room for bakeries—the initial priority—and later for stables, offices, workshops, and all the other activities in this rapidly expanding but determinedly self-sufficient company. By the end of the century Lyons was selling pre-packed tea to shopkeepers all over the country, building on the success of the teashops and the variable quality of existing tea suppliers, who were widely perceived (often rightly) to adulterate their loose tea. One of the earliest brands was Lyons Green Label tea, still available a century later.

In 1909 Lyons opened the Coventry Street Corner House in London WC2. This was a huge complex of restaurants, carefully designed to appeal to the tastes and pockets of a wide range of patrons, from the working classes on an occasional treat, through office girls, to the aristocracy. The initial capacity of 2,000 soon expanded to 5,500, with all meals, unusually, cooked on the premises rather than centrally prepared at Cadby Hall, and there were also a food hall, hairdresser's, shoe-shine parlor and theater booking service.

In the post-World War I period Lyons reached a pinnacle of high-quality mass catering probably unequaled anywhere before or since. The Empire Exhibition opened in 1924 in a huge new exhibition complex at Wembley. It was designed to show the British people something about the many countries that made up the

Empire. J. Lyons & Co. was the sole caterer, regarded as the only company capable of doing the job without using subcontractors. Thirty-three restaurants covered 10 acres of ground with seating for 30,000, serving 8 million meals in the first year, supported by 70 van deliveries a day, and a specially built branch railway to bring in the less perishable goods. Dozens of minutely defined job descriptions broke the operation down into precise tasks, including, for example, six "cheesemen" in the Grill Room, their sole job to supply the waiters with the correct amounts and types of cheeses. It was part of the contract that every item had to be sourced from an Empire country.

Meanwhile Lyons had started staging massive banquets. The most ambitious of them all, for the Masonic Festival in 1925, was another epic piece of organization. Some 7,250 Freemasons assembled in Olympia, sitting down to 1½ miles of tables and served by over 1,250 waitresses. Breakages alone added up to over 3,500 and, as Peter Bird found, "it took the staff 14 hours just to arrange the flowers."

Around this time a major revamp of the teashop waitresses' uniform to suit the decade of the "flapper" was launched with an impressive marketing campaign, and "Nippy" was coined as a nickname for the waitresses as a result of a Lyons staff competition. The new uniform was modeled for the newspapers, the look was embraced by Lyons customers with enthusiasm, and by 1929 a musical comedy called *Nippy* was on the West End stage. It starred Binnie Hale singing the title song with the memorable chorus:

Nip, nip, nippy, rise up Nippy
When she goes off with her tray
We too are carried away
Neat, sweet Nippy
Quick, swift Nippy

Behind the public face of the company was a very well organized, highly centralized operation. Joe Lyons had long realized that the best way to guarantee the quality of the goods in his shops was to produce them centrally and distribute them quickly. Biscuits, cakes, scones, and all the other dozens of lines available at each and every teashop were produced overnight at Cadby Hall and distributed by Lyons's own transport around London in the early hours. Teashops farther afield, all over England and beyond, were served by the vast and reliable railway network. It is a sobering fact that today's highly planned, highly computerized, high-speed transport and distribution network could scarcely match the system used by Lyons a century ago.

This was a huge operation, meticulously organized. It was the era of "scientific management" and Lyons was in the forefront of such developments. Production processes were carefully planned, implemented, and monitored, and Lyons applied the same approach to clerical work. The company's Systems Research Office (later known simply as Organization & Methods, O&M) became very important and innovative. Lyons was one of the first industries to recruit graduates as management trainees, including John Simmons, who reported directly to the company secretary. Simmons in turn hired Raymond Thompson and these two brilliant mathematicians became hugely important to the company's advances in accounting and office systems in general, and the LEO project in particular.

As the demand for exhibition catering declined with the passing of the age of exhibitions, the teashops became more important to Lyons's profitability, but there was a fundamental problem. Their profits came from tiny margins on high turnover. As long as the shops were busy, they could survive on an average of just a farthing a meal. But the economic problems of the 1930s put that model into question, and by the time World War II broke out Lyons had already been considering the drastic step of dispensing with the beloved Nippy and going self-service. As women headed into industry during the early years of the war this became inevitable, and by the end of the war every surviving teashop was self-service. There were 253 at the start of the war, 70 fewer by the end, and nearly all those closures were due to bomb damage—just one of the London teashops survived the war without any damage at all.

Before the war, John Simmons and his colleagues in Systems Research hadn't just been looking at the teashops for ways of saving labor through more efficient working practices. Simmons also wondered whether office machines could do some of the work of the growing army of clerks in Cadby Hall, perhaps revolutionizing it just as machines had long transformed the production line. That's not to say that this army of clerks was inefficient. Before the war, "the company was an extremely efficient organization in the production side and in distribution and in the offices, and to a much greater extent than most organizations the three sides were closely intertwined," recalls David Caminer, who was spotted as a young man by John Simmons and recruited into the Systems Research department as a management trainee in 1936. But World War II put any ideas of automating office systems on hold; Lyons concentrated on keeping its food production going and applying its management methods to running munitions factories, with great success.

After the war, David Caminer returned from military service as manager of Systems Research and started looking afresh at the company's office operations, but "we had to say there was not a lot more we could do with the materials at our disposal." The obvious place to look for new office equipment and ideas was the USA, whose war effort had not been hindered by homeland bombing. By 1947, when Simmons arranged for Raymond Thompson and another senior Lyons manager, Oliver Standingford, to visit America, the company had been more or less cut off from improvements in office equipment for almost a decade.

Thompson and Standingford set out to look at "everything, office machines, layout, nature of desks and chairs, all these things were in their remit," according to David Caminer. The trip, which took place in May and June of 1947, was at first rather a disappointment. Thompson and Standingford felt they saw nothing that improved on Lyons's existing efficiency, and that the widespread use of tabulating machines was often the result of pressure salesmanship (and a willingness to answer any clerical problem by buying more machines) rather than a scientific evaluation of the problem and the best way to tackle it. But the ENIAC was another matter.

Reports had begun to appear in the British press of an "electronic brain" in America that could in a matter of minutes do complex mathematical calculations that took human brains weeks to work out (even with the help of slide rules and log tables). Thompson and Standingford had been encouraged by these reports to schedule a visit to Dr. Herman Goldstine during their trip.

The ENIAC had been moved from the University of Pennsylvania to the Aberdeen Proving Ground months earlier, and was in the process of being recommissioned, so Thompson and Standingford didn't even see it, but it was the development that most excited the two men. Particularly as they learned that a more advanced version, the EDVAC, was being developed, and this would be more compact, reliable, and versatile. And to their amazement their host, Dr. Goldstine, told them there was a similar machine being made in a laboratory back home, at Cambridge University.

By the time Thompson and Standingford got back to England, then a week's journey by boat, Goldstine had already contacted Douglas Hartree, by now professor of mathematical physics at Cambridge, and Dr. Maurice Wilkes, head of the Mathematical Laboratory, and an invitation to go and see the device was on their desks. Within weeks they were viewing the EDSAC that Wilkes was building on his own initiative, assisted, it seemed to them, by only a draftsman and a couple of students. Wilkes's aim wasn't to stretch the new technology to its limits but to build a useful working machine as quickly as possible to serve the university: "There were Nobel Prize winners queuing up to use the facilities," he recalls.

Significantly the Lyons managers' first report expressed bewilderment at Wilkes's lack of resources. That perhaps reflects the fact they had just returned from America. Despite the spartan conditions, Wilkes was making great progress, but the visitors were right—the team was short of money and staff. So on their next visit they suggested a deal whereby Lyons would put up the money in return for the right to make a commercial version of the machine. David Caminer finds the company's readiness to build its own computer unsurprising because "as soon as Lyons found they couldn't get something they would step in and do it themselves. There was enormous confidence in Lyons that we could do anything so we ran our own motor fleet and did our own repairs, we ran our own laundry, we had a state of the art laboratory, and so on. The company took on the largest catering events ever known in this country, an enormous logistical effort but always confident they could do it. So it was quite natural that when it came to the question of a computer we sat down to produce it ourselves and made it work!"

Lyons also loaned Wilkes one of their technicians, an inventive young man called Ernest "Len" Lenaerts. He had been working on a vending machine that would dispense "a sizzling hot sausage" when the appropriate coin was deposited. This ingenious idea used diathermic heating and might well have done a roaring trade on cold and lonely railway stations late at night. But Lyons's judgment that he would be better employed helping to build a computer was probably right.

Before starting to build their own version the Lyons board of directors wanted evidence that the EDSAC would actually work. It came in May 1949, with a phone call to Simmons saying that Wilkes's computer had successfully run a simple mathematical calculation using a program and data stored in the mercury delay-lines. The board gave the formal go-ahead, but they hadn't wasted the intervening period. They had formed a small team to build the Lyons version, and David Caminer had been transferred from Systems Research to managing the applications—how the computer would be applied to office work. Another important decision had already been made. John Simmons, perhaps in a rare moment of whimsy, decided to call the project "LEO," short for Lyons Electronic Office.

There was no electronics engineer within the company, so Lyons published a rather elliptical advertisement that referred only to "a graduate electronics engineer." The advertisement was answered by Cambridge physicist John Pinkerton, who had spent the war as one of the boffins at the government Telecommunications Research Establishment, already knew Maurice Wilkes, knew what Lyons was up to, and guessed what the ad was about. Pinkerton got the job of building LEO, alongside Caminer's applications team, both men reporting to Raymond Thompson, who was in overall charge of LEO activities.

Construction moved quickly once the go-ahead was given. The team was housed in a building in the middle of the now sprawling Cadby Hall complex (Lyons had over the years bought much of the property surrounding the old piano salerooms). Computers were big then, very big. In 1949 the magazine *Popular Mechanics* had peered into its crystal ball and predicted that "computers in the future may weigh no more than 1½ tons." Rooms, not desktops, were where you put computers then.

John Pinkerton's task was to build a version of the EDSAC that would handle business applications with a reasonable level of reliability. Maurice Wilkes recalls that "he built a somewhat re-engineered copy. He had a rule that he didn't change anything in the EDSAC design unless he thoroughly understood why we had done it that way. These were great words of wisdom from a great man. He didn't use the same vacuum tubes as we did, but they were very similar. And the mechanical form of the chassis was different, but circuit-wise it was very much a copy—the logic, what we would now call the architecture, was exactly that of the EDSAC." Wilkes didn't take an active role in this process of adapting his design to Lyons's needs as he was fully occupied in running the Mathematical Laboratory in general, and particularly in building a university computing service round the EDSAC. He had built his machine and now his main interest was programming it. Instead much of the transfer of technology between Cambridge and Cadby Hall seems to have been accomplished through Len Lenaerts, the young ex-RAF technician whom Lyons had seconded to Wilkes's team along with their initial injection of funds. According to David Caminer, Lenaerts's "service orderliness and his firm determination to get on top of a subject that was largely new to him, together with his natural inventiveness, all contributed to the [Cambridge] project. He kept Cadby Hall in touch with frequent reports, giving a clear exposition of the almost undocumented work that was going on."

Pinkerton's friendship with Wilkes, and his frequent visits to Cambridge, must have helped the process too. His team already knew they would need extra mercury delay-lines to store more data than the EDSAC needed. More importantly they knew that the input and output of data would be a much more demanding task for LEO than for the EDSAC: "In science the whole emphasis is on massive calculations based on little data and producing little in the way of results," says Caminer. To work out the path of a shell fired from a gun, you need little more data than the weight and drag of the shell and the angle of firing. It is a complex calculation that might take a skilled mathematician days to calculate, yet it produces just distance and height as the result. But "for clerical work it was quite different, with masses of data and masses of results. So we had to introduce something like a three-ring circus, with the data for one person being taken in at the same time as the calculation was being carried out for the person before and the results were being printed for the person before that. And equally several lines of input were coming in concurrently and several lines of output were going out concurrently."

Their preferred solution was to use magnetic tape to read the data into the computer and to output the results quickly enough to keep pace with the electronic speed of the calculations. Later on, magnetic tape would be a mainstay of computer data input and output but at that time the technology was too new (and across the Atlantic the UNIVAC team was facing similar problems with their phosphor-bronze tape machines). Lyons engaged a company that was doing some telephone installation work at Cadby Hall to design such a machine, but this was one aspect of the project that went badly and time was lost waiting in vain for it to work.

While the rest of the mechanical development proceeded, the team also had to think about how they would program the computer. In the early post-war years a company couldn't simply advertise for a computer programmer. There were no business programming courses; indeed there were scarcely any programming languages, so Lyons again looked to its own resources, searching its own staff for people who might have an aptitude for programming.

The first recruit was Derek Hemy, another management trainee who had finished the war in Signals Intelligence and was now a member of Systems Research. He was sent off to Cambridge to participate in the work of Maurice Wilkes's team in developing programming as a technique. An early trawl also brought up John Grover, yet another management trainee and a Sword of Honor pilot from the RAF. As a hands-on supervisor he standardized and enforced the new programming techniques that were being developed. Then came Leo Fantl from the air force and Tony Barnes from the navy.

It was time now to throw the net wider, so the door was opened to anyone in Lyons who thought that they might be capable of programming LEO. David Caminer designed a course to identify the likelier candidates and as this too was new territory he devised the tests from scratch. One of the people found this way was Frank Land, a graduate of the London School of Economics who had joined Lyons as a clerk in the statistics office. The tests were part of what Land remembers as "a very intensive course and each night I would go home and, with my wife, who was also an academic, work out what we had to do and with her help I managed to pass that course all right. The course was very much concerned with how a computer worked but above all in getting the logical steps right in defining what a computer had to do."

Another find was Mary Blood, daughter of the company's chief medical officer, whom Caminer remembers as "an extraordinarily formidable figure, a frightening man. He asked for her to be given a trial and we took her on. She had a languages degree." More than a decade later it was still believed that arts graduates made the best programmers, though Mary Blood was also "fairly mathematically minded." She and Frank Land went on the same four-day course and were the only two candidates immediately recruited to the LEO project. She says the course was designed "to see if you had the right sort of reasoning skills to break problems down to their basic elements. We had a test paper where we had to work out the answers to questions that were simple but tricky. One question was something like: write down the instructions in the smallest detail to work out the difference between ¾ and ¼."

Program development was immensely complicated. The computer broke down frequently in the early days, and even when it worked programs would fail many times in development. There was no way of simulating a program; the only test was to try it. As Frank Land recalls, "It was hands on. We fed it into the machine and we diagnosed what was wrong. I remember you had a log in which you entered the starting time and so on, and oh the joy when you had the entry in the log 'passed point of previous stoppage,' indicating we had made progress!"

David Caminer says the team was "laughably small" by the standards of today, and Frank Land recalls, "On the technical engineering side there were maybe 10 or 12 people; on the programming and systems side probably six or seven." They were also "rather special, perhaps not from the ordinary material! One of the ways you could see this was in the cars they drove. They arrived here at Elms House [on the edge of the Cadby Hall complex] in a variety of vehicles. There was a guy who came in a Rolls-Royce, a very ancient one. I had a 1933 London taxi, one which you could open from the back. There was the Messerschmitt which Payman had— I remember when Payman and Jacobs went to Spain in the Messerschmitt—it was a little three-wheeler, a bubble-car. There were quite a few Vespa-type things as well, and a Morgan three-wheeler. There was an Alvis. . . . The people were quite eccentric perhaps."

While he waited for the magnetic-tape problem to be resolved, Caminer pushed to get at least one application running on the partly built LEO. "I invented this job of bakery valuations, valuing all the goods that came out of the Lyons bakeries." This information told managers what was really happening in their bakery production, and it was normally calculated by hand and human brain. It was a good first application because it didn't either need or produce large amounts of data. For this reason Simmons thought it rather pointless to computerize the task, but Caminer responded, "Yes, but we do need experience of getting live work done, and done to time. So Simmons reluctantly agreed and that's how the world's first business application came to be born. We had a few trial weeks, then in November 1951 the job went live, and carried on every week afterwards."

Payroll was the first of the real applications that would justify the time and expenditure on LEO. When it became clear that the magnetic-tape input devices just weren't going to work, the team fell back on their alternative plan to use a combination of paper tape and punched card. The brilliant John Pinkerton connected them and this allowed the rest of the team to get on with developing the payroll application. If Pinkerton's role in the LEO story has sometimes been under-reported, it's because, in Maurice Wilkes's words, "he did such a good job that people just took it for granted," which is about the highest compliment an engineer can hope for, though unhelpful to public recognition.

It is said that when the revered company secretary of Lyons, George Booth, asked Pinkerton at his interview, "Do you think you can build this, young man?" he replied, "I think so but I don't think it will be very reliable." That was one of David Caminer's main worries too. The problem was the valves, thousands of which would be needed inside LEO. Each valve was a vacuum-filled glass bulb with electrodes and a heater inside and connecting pins outside. They were the best technology of the time, but heaters burned out, glass cracked, connecting pins made bad contact, and so on. A single valve failure would stop the computer or at least produce errors in calculation.

To cope with this the engineering team, says Caminer, "went to enormous trouble to try them out. Before even fitting them they 'aged' them to expose any frailty. When they were in the machine we carried out exacting test programs before ever starting on a real job, trying the whole circuit with lower voltage and higher voltage to see if they were susceptible. We even went to the extent of vibrating the racks from time to time." It worked. "Breakdowns happened but it was our job to manage round them. We were never prepared to say we couldn't deliver the goods because the computer was down. It just never occurred, and that's funny as it's an excuse that's trotted out so regularly now!"

Caminer's background in Systems Research was very useful as it was "really rather simple for us to do the payroll applications. We knew how they were carried out and what purpose they served. So we took what we knew and added on anything conceivably useful, like at the end of the calculation we rounded off pay to half-crowns to save a lot of handling of coins   all with the agreement of the workforce of course. It really was a model job and when Pinkerton and I went to the States four years later we found nothing nearly so advanced, in the way the machine covered the whole scope from the clock card to the pay packet and all the ancillary calculations." Other US companies with European operations, like Ford and Kodak, soon heard about this and paid to have their payrolls calculated on the first LEO.

Next came LEO's most striking application of all. "From the earliest days of LEO's development," says Caminer, "the application which struck us was the replenishment of teashops every day. As the system stood, there was a formidable manageress in every shop and she sat down every afternoon with pre-printed order pads and these were couriered into Cadby Hall and then the delivery notes made out. I looked at that and it was clear to me that there was too much data to be able to record it and get the computer to do it in time. So we had to make a change if we were to succeed in using a computer."

This was one of the secrets of the LEO team's success. They didn't simply reproduce existing clerical procedures in the computer.

Caminer arranged with Lyons's reconstituted Systems Research office to look at the whole process: "I had heaps of these orders on my desk, trying to make some pattern out of them. What I didn't know was there already was a simple pattern—patterns generally are simple." What he spotted was that while the orders changed from day to day, they didn't change much from week to week for a particular day. So a manageress would order much the same things every Tuesday, but a different mix on Wednesdays. "They were not conscious they were doing it, but they were obviously looking up what they'd done the week before. So we arranged that each manageress would set a standard order for each day of the week, and every day she would look at that and see if she wanted to alter it. So all we had each day was a comparatively limited set of alterations."

As a solution it echoed the way the team handled the Lyons payroll on LEO: each employee had a punched card full of the personal information that didn't change from week to week, such as their staff number, basic pay, and so on, and a second card giving only the special data for that particular week (time off, overtime, bonuses, etc.). When an employee's conditions changed, through promotion for example, a new master card was punched. In the same way if a standard teashop order needed a long-term change, a new card would be prepared for it.

To handle the daily amendments to the standard orders, Caminer set up an in-house call center where there were "young women with headsets and card punches in front of them putting in the alterations, which minimized the data entry requirements." Yvonne Dolezal was one of them: "I was young, and to be in at the birth was very exciting. There was loads of hubbub, meetings all the time, doing trial runs, all to get that first successful program. You had your earphones on and when someone came on the line you would ask them which shop they were calling from. Each shop had a number and the manageress, or sometimes the deputy, would say 'Oxford Street' or whatever. You typed in their name and number and then they would tell you what they wanted to order. Every item had a code number, like cupcakes might be 102, and if they wanted four trays you typed '102 × 4.' And to know which operator had typed which shop's order you typed in your number. Each operator had her own number that you stamped on every card before you started. I was number 10. I've never forgotten."

While morning shifts at the call center were fairly predictable, from 8 a.m. to 2 p.m., evening shifts were not. Once the cards were punched for all the shops' amendments that day, they were taken into the computer room and fed into the card readers. "After we'd finished in the punch room, we had to sit around until the printout came off. There were loads of pages for each shop, and we broke them down via a crib into shops. So a huge stack of printout had to be cut on the old guillotine. Then it had to be clipped to a clipboard. That would take a couple of hours, and we might get away by 9 or 10 o'clock. But if LEO broke down they had the floor up, they had the sides out—it could take hours and hours. Didn't happen that often though. I've known it to happen on Christmas Day, on bank holidays, don't know why . . . maybe they were more merry and not quite with it! But it was very interesting to see how they tackled the repairs."

Once the printouts were sorted they were taken into the factory and the girls could go home. While they slept, the bakeries would churn out hundreds of different items in the precise quantities dictated by the LEO printouts and by the early hours of the morning they were ready to load the delivery vans. Still LEO's work wasn't over, as the printouts included the delivery notes for each individual teashop. "The computer produced the delivery notes in exactly the right order so that the last shop to be delivered to had its goods put in the van first and the first shop had its goods last. So there was no burrowing about, the tray was there. All the time we were trying to do that kind of thing to make delivery easier, as well as coming up with the figures for the accountants. We were thinking about the whole operation," says Caminer.

The teashops distribution program started running in October 1954, and within days comments such as "We would like to give thanks for LEO. . . . This is a wonderful timesaver and we are grateful for it" were coming back from manageresses, who had never come across a computer before. Just as David Caminer had intended, the new system where only changes to the standard order needed to be phoned in each day "vastly reduced the amount of time the manageresses had to spend at their desks. While they were at their desks they couldn't be supervising their staff, they couldn't be cheerful to customers, they couldn't be advancing their trade, and so on. So one of the things that pleased us most was when we did receive thanks from the manageresses, as we didn't know whether we were going to satisfy them. We knew how very competent they were; we knew how they had the ear of the board of the company too. So when they were satisfied we were exceedingly happy!"

To Caminer's eternal regret the one group whose working methods they failed to change was senior management. "We tried to give the directors and management more information than they had ever had before. Previously they tended to go through reams of paper to see what had happened. Well, the computer was perfectly capable of doing that very much better than they could. And so we did produce a number of statistics for management attention each week—information that we would have wanted if we'd been running the teashops—but we simply couldn't convince the management that this was the way to proceed. They still wanted printouts so they could go through the figures themselves and I fear that's what is still happening today. I think that top management is going through figures on screens looking for information that again the computer could give them much more incisively."

The impact of the payroll and teashop programs in the early 1950s was considerable. No other UK company was running office applications on computer and even before the LEO team had built a second machine they began to win business from other big names. As well as the Ford, Kodak, and other payrolls, there was a more unusual request from British Railways. An act of Parliament had decreed that in the future all rail freight charges were to be calculated on the basis of the actual distance between the start and finish stations, and the legislation had been passed without anyone in Parliament considering whether it was a practical proposition to make these calculations.

To charge freight in this way you had to know the distance between every single pair of over 5,000 railway stations. Luckily the British Railways manager tasked with producing the new freight charging tables had heard about LEO. Caminer recalls: "A little man came from St. Pancras one day. He'd found it quite impossible to produce the tables in the time available before the legislation took effect. So we thought it out, and it was a lovely puzzle. It still took an enormous time to do, seven days a week, 24 hours working, but we did produce their figures in time to take effect." The solution to the "lovely puzzle" was to "break the country up into manageable areas. The next step was to calculate all the distances within those areas. Then the distances were calculated between stations in one area and those in another. Finally, the program chose which was the shortest of several possible routes between pairs of stations." Breaking the problem down in this way enabled Caminer's team to work around the disruptions caused by LEO stoppages, which were still quite frequent. He also insisted that the program calculate distances in both directions, B to A as well as A to B, and compare the results as an error-check.

A job that really should have launched LEO into the public eye was the calculation of tax tables for the government. The annual changes to tax rates in the Budget meant a scramble to calculate the new tax tables, not just for the standard rates of income tax but also for all the obscure special cases, such as merchant seamen. In 1954 the Treasury called on LEO. This was one of Frank Land's early tasks. "Of course the Budget was highly secret, so we had to make these programs very flexible indeed to cover almost any change you could conceive in the way taxation was planned. On the day of the Budget the Chancellor made his announcement in the House of Commons and then we had to wait in Elms House here for the courier to come from the House of Commons, from the Treasury, with the parameters for the new tax tables."

The first year was an anticlimax as the Chancellor made no changes at all. But the next year everything went to plan: "As soon as they arrived we would feed those parameters onto punched tape, feed it into the machine, and start printing the tax tables. And, of course, we hoped there was nothing in the changes in tax we hadn't [accounted] for."

Regrettably the following year the government found it could do the job more cheaply on a subsidized research computer. Official lack of support for LEO's pioneering efforts still rankles with Caminer: "It was an example of the sheer short-sightedness there was then. Right through this period we had minimal government support. They simply didn't realize that business computing would become vastly more important in volume than scientific computing. If they could find some scientific computer with time to spare to do the tax tables, then they went there if they were saving a few bob. It was very sad."

A few astute commentators proved more far-sighted, like the noted economist Mary Goldring, who has been applying her critical faculties to British industry since the late 1940s. Writing in *The Economist* early in 1954, when LEO was still little known outside Lyons, she asked if this was "the first step in an accounting revolution or merely an interesting and expensive experiment." She went on to identify three groups in industry: those who didn't believe electronics had any place in business, those who thought there were limited possibilities, and finally "a third group—of whom Lyons is one—who consider that a major revolution in office methods may be possible."

Three years later Lyons produced an impressive publicity film that made its point in the opening words: "Despite the large number of clerks in modern offices, sufficient numbers are hard to find. To fulfil this modern need came LEO, the first computer built for office work." It went on to show some of the variety of applications that had been achieved by that time, including the teashops distribution. It must have seemed like science fiction to many in business at that time, yet by then the LEO II was available to buy.

John Simmons had always intended that a copy of LEO would be built as a backup. Lyons management could be bold but they were also prudent and no one wanted to find the wages couldn't be paid because the computer was down. Although the early manual backup system was never needed—it's a matter of pride to Caminer that they never failed to get the payroll out on time, even with only one machine—still a backup machine became even more important as the company started taking on outside work. It was supposed to be a simple copy of the first LEO but inevitably the engineers took the opportunity to make many detailed improvements, though the LEO II was still based on valve technology.

By now the whole LEO project had moved to Elms House, a squat late-1930s concrete-and-glass building on the edge of the Cadby Hall complex. The building is still there, now inhabited by

EMI and with no indication of its illustrious part in Britain's computing history. On a brief return visit in 2001, Frank Land remembered it with affection: "We occupied an open office that we called the 'goldfish bowl,' as anyone going past could see us. David Caminer had his own office, but the rest of us were together in the goldfish bowl. We worked very long hours, often at night, when we had better access to the machine to test our programs. And late at night we had the privilege of eating in the senior dining room at Cadby Hall and that was really great; the food was first rate. Lyons was an interesting company. It had on the one hand social cohesion, a sports club, and people were members of the family, but at the same time it was clearly partitioned between different levels. So lavatories were separated into 'managers' and 'others,' and there were different canteens for different grades of staff, so as you advanced up the line you got keys to the lavatories and the distinction of being able to go to a better dining room!"

It was programming the computer that really captured Frank Land's enthusiasm: "It was very exciting because everything you did had never been done before. There was very little repetition and it was a challenge because there were things you didn't quite know how to tackle. You had guidance from the people like Derek Hemy, who subsequently joined EMI, and John Grover and Leo Fantl. They were the ones who had some experience. There was a buzz about the place! At mealtimes you would be discussing what you'd done, any new wrinkles, any new ideas we'd got, and I certainly admired the skill of people like Derek Hemy. The first job I got was checking some code he had written. He was an outstanding programmer and the complexity of the code was unbelievable to me. It took me a long time to understand what he was trying to do, let alone check what he was doing, but it taught me a lot. Gradually I got more familiar with programming."

LEO II was more powerful and more reliable than the first LEO and 11 were built. It had long been apparent that there was a market for LEO computers, not just for renting time on their in-house machine but for selling complete systems to other companies. At the same time as authorizing LEO II in mid-1954, the company publicly stated that it was willing to build additional ones for sale or hire. Lyons was already concerned that potential customers would be put off buying from a company that had no record of making electronic machines or even office equipment, so at the same time they decided to form a subsidiary company. Later the same year, LEO Computers Ltd. was formed.

With LEO II they had a design they could manufacture, although still in small quantities. This rather suited the team's philosophy, which was not just to sell a computer but to provide a complete solution to their customers' needs. Often this would mean changing the way a company worked to get the most out of the application. So, for example, a steel-making company, Stewarts & Lloyds, had already bought a LEO II for payroll and pipe-stressing calculations. Once those applications were successfully computerized the company asked the LEO team about a more unusual problem. They had a regular job of working out the best place to dig for iron ore the next day, and this was carried out by engineers who worked more from instinct and long experience than by following written rules. Nonetheless the LEO people found they could simulate the process in a computer program and what David Caminer calls "this fascinating job" was transferred to LEO, easing Stewarts & Lloyds's worries over who would eventually replace their aging experts.

Caminer can remember only one occasion when this close involvement in a customer's application failed, and that was due to the client company embarking on a major restructuring without any warning to the LEO engineers. However, there were downsides for the LEO company in this approach. It normally meant seconding one of the team to the client for weeks or months as they reworked the process. Often the client company ended up recruiting the LEO specialist, weakening the team (it was still hard to find computer specialists). There was a tendency to undercharge for the amount of work that LEO was putting in, yet even then the overall price could look higher than those of their rivals. When other firms started offering off-the-shelf systems, with exaggerated claims about their ability to tackle any problem, they were inevitably cheaper, on paper. It also meant that LEO was geared up to providing only small numbers of computers, where IBM in particular was thinking in terms of hundreds, then thousands (although IBM had come late to electronic computing, it envisaged production on the same scale as its other office equipment). Almost every LEO II differed from the previous one, because of the engineers' constant desire to keep improving the design, and that too was a rather uncommercial practice.

By the end of the 1950s LEO II was looking dated and work started on the next generation. LEO III would use transistors and other semiconductor components instead of valves. It would have microprogramming, a technique invented by Maurice Wilkes that made the hardware design easier and more reliable. It would have multiprogramming, allowing several programs to run at the same time, and its arithmetic operations would run about 10 times faster than those of LEO II. When it launched, rather quietly, in 1961 it was, says Caminer, "three or four years ahead of the IBM 360 and better in many respects." Within the LEO III family there were two important variants, more powerful machines known as the LEO 326 and the LEO 360.

By now LEO Computers had moved again, out of Elms House and into more central premises above Whiteleys in Bayswater (the world's first department store, now a shopping center), which was named Hartree House in honor of the Cambridge professor who had played such an important part in the process that led to LEO's predecessor, the EDSAC, developed by Maurice Wilkes. The move away from the Cadby Hall area was another step in creating a separate identity for LEO Computers Ltd.

LEO III was everything the company had hoped for in terms of its performance and it was much the most successful version of LEO commercially, with about 80 built and sold to customers outside Lyons. A fair proportion of these were overseas, some behind the Iron Curtain, while others went to South Africa and Australia. Frank Land's twin brother, Ralph, had been a management accountant in the teashops division for some years before going into LEO Sales, where he found himself handling a lot of the business with the communist states: "Many Eastern European countries had LEO in the key organs of the state." The original concept of selling behind the Iron Curtain was developed by Dan Broido, whose parents were born in Russia, while he was born in exile in Irkutsk. After the Bolshevik revolution they emigrated to Germany; then when Hitler came to power they moved to the UK. "Broido had the idea one could sell computers in those countries," says Ralph Land, "and I was given responsibility for this. We developed quite a significant business, and with a smaller staff than IBM we still kept market leadership over them!"

It was sometimes a surreal experience: "We were dealing with foreign trade organizations who knew nothing about computers. The process of getting authority to buy was immensely complicated. They had maybe 15 organizations having to sign off authority to buy computers, including the secret police." They sold to a range of Comecon countries: Czechoslovakia, Poland, Romania, Bulgaria, the Soviet Union, and Yugoslavia. Though the officials may have known little about computers, they were trained and highly skilled negotiators: "We had to find ways of generating finance for our customers, for example by doing barter deals, by helping them to sell products in order to raise the foreign currency to buy computers. So it was a very different business under difficult circumstances. Giving out economic information could result in the death penalty, as it was giving away state secrets! If we wanted to estimate the storage details for a computer, we needed to know them, but the details we were given were often quite fraudulent. We had to learn to cope with all that. If they became too friendly we knew they were secret service, but we established good relationships, and we went through the Prague Spring in 1968. I think it helped the process of opening up those countries."

Back home the best customer for LEO III was the Post Office, at the time one of the biggest users of engineering in the country and "certainly one of the most capable O&M departments—and largely independent of the Civil Service, because it was so competent," recalls David Caminer. "They quickly cottoned onto the capability of LEO III, which they saw as the answer to their needs. Over several years we put many outstanding applications on to their systems and installed a network of LEO 326s all around the country. The telephone billing job was the biggest billing job in the world at that time. We built the Premium Bonds business with them—that was another job where legislation said it had to start on a certain date. We built the Girobank for them, in Bootle—very close to Harold Wilson's heart!" Bootle is on Merseyside and Harold Wilson, then the Prime Minister and the MP for nearby Huyton, officially opened the National Girobank in 1968.

Unfortunately there were some companies that just couldn't bring themselves to consider buying a computer from the same business that made their Swiss rolls. It wasn't a problem with clients like Imperial Tobacco and Stewarts & Lloyds, "as they knew Simmons and they had some confidence in him," says Caminer, "but the outside world, my gracious yes! I remember going to Imperial

Chemicals, as it was in those days [today it's ICI], and talking to an old O&M acquaintance. Clearly it was out of the question for him to ask a catering company to do any technical work."

Frank Land agrees there was an image problem but thinks "other companies like IBM were also outselling us simply by their technique and marketing. We were not in the same league. I said earlier how confident Lyons was but this was an area they didn't grasp. There was no question of it, people were more likely to go to IBM, and IBM was big! An executive who chose IBM could never be faulted but Lyons was felt to be a risk." Moreover, it wasn't just technical companies that felt this way; "even inside the government machine there was a certain amount of doubt. It was felt that support had to be given to the electronic companies, Ferranti, English Electric, AEI, EMI, and so on. It has to be realized there were more companies trying to build computers in Britain [by the 1960s] than in the whole of the rest of Europe and probably as many as in the United States. The fragmentation here was unbelievable and the government found itself powerless to bring anything together."

By the early 1960s LEO Computers Ltd. was under pressure from several different directions and Lyons senior management felt it couldn't go on underwriting the expansion of LEO development and sales. So they went in search of a partner for LEO and in 1963 it was merged with English Electric, which was then one of Britain's largest engineering firms. The new company was called English Electric LEO and Lyons retained a stake in it.

However, English Electric was already in the business computer market, building a design licensed from the American company RCA, "so there was a conflict between the LEO people with their LEO III and the English Electric people with their machine, the KDP10, which was really rather clapped out by that time. The two teams never really got to work properly together at all and that's quite common in computers. People get to love what they know well and find it difficult to see any virtue in something that comes from elsewhere. So that wasn't a very happy merger," in Caminer's judgment.

Moreover, the English Electric staff had a different approach to selling. The LEO team still wanted to work closely with customers to get their applications working as smoothly and efficiently as possible, while the English Electric people "saw their job as selling machines to customers who would make them work. The difference was enormous, so it wasn't a happy relationship." The English Electric approach was in line with trends in purchasing, as clients wanted suppliers to compete against each other to provide the best tender against a set specification. In vain the experienced LEO installers argued that this wasn't the best way to approach a major computing job.

Lyons sold its remaining stake to Marconi in 1964 and the company became English Electric LEO Marconi. Three years later, when English Electric LEO Marconi bought Elliot Automation, it was decided there were more than enough brands in the name already and the company rechristened itself English Electric Computers. The same year the last LEO 326s were being installed, one of them ironically at Lyons, and it was time for a new model. English Electric Computers had to choose between developing a highly advanced in-house LEO IV, sketched out by John Pinkerton, or adapting the latest RCA range. The expense of the former approach couldn't be justified and the LEO name disappeared, although they did call it System 4. Says Caminer, "They carried LEO thinking right into it. We changed the software completely, but there was a break in the tradition. So the LEO name had gone out of the company and now it went out of the system as well." While the brand disappeared from both the product range and the company name, there was plenty of life left in many of the installed

LEO computers (the Post Office LEO 326s would remain in use for more than 14 years).

Only two years earlier, in 1965, the very first LEO had been switched off for the last time, earning a tribute and editorial in the *Daily Mail*: "Today we mourn the passing of a computer ... throughout almost 14 years of life he worked a 24-hour shift on one dreary problem after another without complaining and spent, at most, only a few hours off sick."

CHAPTER 6

SO THEN WE TOOK THE ROOF OFF

The building at Feofania on the outskirts of Kyiv, Ukraine, where the first Soviet electronic computer was built. Originally a monastery, it served as a mental asylum, a military hospital, and a computer laboratory and is now restored as a convent (author's photo).

A few miles southwest of Kyiv lies Feofania, set on the side of a hill among long-established trees. It has a fine church—some call it a cathedral—sitting on a gentle rise, its golden steeple piercing high into a blue sky, overlooking a small lake. Along the side of the road that winds down through Feofania are a few houses, including a larger, rather squat building, a convent that started life as a monastery. The church is now a popular place for weddings and on weekends family groups in their best clothes pose outside for photographs. Occasionally the priest can be seen blessing a family's car, still a prized possession as Ukrainians try to catch up with the Western economic miracle that they could only glimpse for so many years. Maybe religious veneration of the motor car says something else about the Ukrainian character too, hinting at the high regard they have for things made by people. A hint reinforced by the name of the road, Academician Lebedev Street.

Sixty years ago this area had been laid waste, like much of war-torn Ukraine. Informed estimates suggest at least 7.5 million Ukrainians died in World War II, more than the total military losses of the USA, Canada, the British Commonwealth, France, Germany, and Italy put together. It was the major battleground on the eastern front. The German Wehrmacht swept across from the west in 1941, slaughtering and pillaging as they advanced, and occupying Kyiv on September 10. They retreated just over two years later, destroying as much as possible to delay the Red Army counterattack. By the time the country was "liberated," there was little to celebrate other than liberation itself. One-third of the population lay dead and millions more were in camps in Germany, many of the survivors eventually migrating as "displaced persons" to Britain, America, and other countries.

Ukraine's 1.5 million Jews suffered in particular, with 600,000 exterminated. So too did the mentally ill. The monks had left Feofania's monastery long before the war, during the religious suppression that followed the October Revolution of 1917, and the building became a mental asylum. When the Nazis arrived they killed all the inmates; they had little regard for the *untermensch*, the "subhuman" peoples of the East, still less for their mentally ill. They turned the building into a military hospital, and both it and the church were badly damaged in the final battle for Kyiv. By the end of the war the capital city itself was a ruin, says Viktor Ivanenko, with half its houses wrecked and much of its pre-war industry destroyed or transferred deep into Russia.

Ivanenko remembers it very well. He was a young man of 14 working in a garage when the German army overran Kyiv. There were partisans among his older colleagues and he was imprisoned along with them. For a while they were guarded by Ukrainian collaborators in a prison camp; then they were put on cattle trucks for Germany. Viktor was lucky, escaping from the train, returning to Kyiv, and surviving the war. "The central part of the city was destroyed and everyone lived in very difficult circumstances, very overcrowded apartments, and workplaces were also crowded." When peace returned he became a student, eventually qualifying as an engineer and pursuing an academic career that he still maintains even in his nominal retirement.

Out of the chaos of the early post-war years came the first Soviet computer, a story that until recently was almost unknown to the Western world. Its creation was largely due to an engineering genius, Sergei Lebedev, born in Siberia on November 2, 1902, and the man whose name adorns the road through Feofania. When Lebedev was at university in Moscow in the late 1920s, electrical engineering was popular and highly regarded and that's what he studied. In his spare time he also became a keen and talented mountaineer, as did several of his colleagues. By the mid-1930s he had his doctorate for a thesis on the stability of main power supply systems. During that decade he was involved with early analog computers for the modeling of power supply distribution networks and the solution of differential equations.

This was important work in the Soviet Union because of the drive to industrialize and in particular the need for a reliable national electricity grid. The mathematics and the engineering involved in such a system are formidable and this was one of Lebedev's interests. The Russian historian Igor Apokin claims that as early as the late 1930s Lebedev was working on a project for an electronic computer using binary arithmetic, but the work was terminated by the outbreak of war in 1941. Certainly his widow recalled much later that during the early years of the war he spent many hours covering sheets of paper with 1s and 0s—binary arithmetic—working by the light of a candle.

These memories are quite plausible. While it was a comfortable assumption during the Cold War that the Soviet Union simply copied its technology from the West, this was not always true. In electronics the "flip-flop," a crucial component of the early electronic computers, was invented independently in Russia (by Mikhail Bonch-Bruyevich) and in America (by William Eccles and F. W. Jordan). Indeed the Russian invention in 1917 preceded the American by two years and Bonch-Bruyevich went on to devise the first water-cooled electronic valve, an extraordinary device that had as many of the metal parts as possible mounted outside the glass vacuum bulb, which was immersed in running water. This strange design enabled the device to handle much more power than any valve before it, and the two inventions together enabled the Russians to claim the honor in 1921 of the first live radio broadcast with a human voice (instead of Morse code, which was the most that lower-powered broadcasting could manage before then).

Likewise the semiconductor p–n junction, the heart of the transistor, was created in a laboratory in 1941 by the Ukrainian physicist Vadim Lashkarev, though he didn't go on to develop working transistors until the early 1950s. This was later than the Americans John Bardeen, Walter Brattain, and William Shockley, who in 1947 independently created the world's first transistors, for which they later won a Nobel Prize. They had no knowledge then of Lashkarev's work. The development of electronic technology in the 1920s, 1930s, and 1940s was an international phenomenon, with no one country responsible for everything. The world's first scanner (or "electro-optical reading machine") was unveiled in Moscow in 1938 and there were other examples of similar advances made independently, as part of the natural progression of technology.

The Soviet Union was in too much trouble to work on computers during the war, and they certainly weren't going to start any such project in the Ukraine. On the contrary they pulled as much out of the country as they could; one of the great feats of civil society under conditions of "total war" was the mass movement of heavy industry from the Ukraine to the east of the Ural Mountains, which in effect separate Europe from Asia, before the advancing German army.

However, as post-war reconstruction got under way, Sergei Lebedev's interest was kindled anew by reports of "electronic brains" coming out of the West. There was little technical information accompanying these reports, and little espionage to speak of. But it didn't matter. Once the idea of electronic computing was established as a practical proposition, Lebedev and people like him could work out what to do. A much bigger problem was the electronic and associated mechanical components. Both in quality and in quantity the USSR lagged behind the West, but they too had been spurred on by the rapid development of radar and military communications during the war, so they had enough to work with.

They also had a powerful incentive, the most powerful force known to humankind. When America detonated atom bombs over the cities of Hiroshima and Nagasaki, it brought the war with Japan to an end, but it also sent a message to the Soviet leadership. Arguments have raged over those bombings ever since, but there's little doubt the American government meant the message to reach the Kremlin and equally little doubt that it petrified the USSR. With an uneasy, heavily armed, peace in central Europe, and Western talk of a follow-on war with Russia, Russia believed it needed the atom bomb. That in turn meant having a computer to solve the complex equations involved in its development and manufacture.

So in 1946 Lebedev was persuaded to move with his young family from Moscow to Kyiv to become the director of the Institute of Electrical Engineering. There he set up a special seminar with scientists from a range of fields and they discussed the sort of problems that could only be solved by a computer and how it might be designed. He was elected to the Ukrainian National Academy of Sciences and given his own Laboratory of Automatic Control. Late in 1947 he started work on his *Malaya Elektronnaya Schetnaya Mashina*, or "Small Electronic Calculating Machine," always subsequently referred to simply as MESM (pronounced "MEZZ-im"). Originally the first word was "Model" (the same in English as in Russian), but he was advised that a "model" calculator would not be taken seriously.

One of the men who joined him was Zinovy Rabinovich, at the time a gifted young engineer studying for his doctorate. Later Rabinovich forged a distinguished career as a professor, and in retirement he remains a regular visitor to the Ukrainian Academy of Sciences building in Kyiv. There he recalls ruefully his reaction to Lebedev's plan to build a computer. "We were working on automatic control of electrical systems and particularly space systems and so my dissertation concerned this topic. Then in August 1948 Lebedev announced that our lab would work on another program, on computer science. I personally was astonished because I thought we were working on such advanced theory and now we would work on this *calculating* machine. And then Lebedev told me that I didn't understand anything! He explained to me that computer science was getting very important now, as important as nuclear power."

It wasn't just his staff who harbored doubts; Lebedev was also acting in the face of scorn from his superiors. Rabinovich says "at this time the academy did not accept this work so MESM was made with money received from the Institute of Rocketry. Lebedev's work on computers was recognized only when people saw that something very great was being done."

At that time anyone developing a computer under the Soviet system faced another problem, that of the "ideological environment," which rejected the whole field of cybernetics, with its implicit comparison of human and machine behavior, as a pseudo-science. Opinions vary on whether this inhibited the development of Soviet computing, Igor Apokin arguing that it hampered the study of social and economic modeling but not computing, which he says was never regarded as reactionary or anti-scientific. Nonetheless several years later, when Lebedev gave a presentation on the completed MESM to the Soviet Academy of Sciences in Moscow, he was criticized for using the term "logic" to describe electronic circuits, on the ideological grounds that "logic" was a characteristic of humans not machines.

Still Lebedev got the support and resources necessary to build MESM. He needed a spacious building and when he found Feofania it had the added attraction of being outside Kyiv, making security easier. The church was a hollow shell; the former monastery next door was also now empty and half-destroyed. Both buildings were owned by the Ukrainian Academy of Sciences, which had already allowed the Institute of Mechanics to put the church to a rather more worldly use. From inside the top of the tall spire they had at least 20 meters to hang lift cables designed for mineshaft work and test them to destruction. From the very beginning they were surrounded by guards, with a policemen always on duty at the entrance, and the building was named Secret Laboratory Number 1 (though another account has the less enticing title Laboratory of Modeling and Adjustment, which would have been a better way to keep a secret).

The academy had no particular use for the ruined monastery and handed it over to Lebedev's team, who had to build their laboratory and workshops from scratch. But it was a welcome offer given the state of Kyiv at the time, according to Viktor Ivanenko. "Here in Feofania the team had much better conditions, so it was a very good decision of the Academy of Sciences to put the team here. And it was Lebedev's choice; he found this place."

Another engineer, Rostislav Cherniak, joined the team in 1947, and was there when they started work on the two-story building the following year: "The monastery was in ruins, so we rebuilt it and we also built an electrical power station. It gave 6 kilovolts and we even gave power to an observatory the Academy of Sciences was building nearby. We had to do everything, even build a metalworking shop. That was all on the first floor (at ground level). To the left was the laboratory and a bigger room where we assembled the computer."

It sounds like hard labor but they counted themselves very fortunate. Although the buildings were in a poor state, the area was as idyllic then as now. Most of the team lived in the settlement, upstairs in the same building that housed MESM. Cherniak's daughter Svetlana was seven when she moved there with her family. "The people on the team were very different from other people," she recalls. "They were like people of the twenty-first century. They were very friendly, they loved each other, they worked night and day without counting the hours. Most of them lived on the third floor in small apartments that were really just rooms. Lebedev traveled from Kyiv and later some of the members of the team like Malinovsky also commuted. But most lived in Feofania. We children remember it as a very good time because Lebedev's family had four children: his wife Alisa had three, Sergei, Natasha, and Katya, and they also adopted a small orphaned boy. When they had breaks, the adults played volleyball, swam in the lake, and spent time with the children. We had no formal education but we followed the adults. They took us to the forests to collect mushrooms. There was such a warm feeling of community and at the time it just seemed normal, but now I realize it was exceptional."

One of the few other survivors of that team is Boris Nikolayevich Malinovsky, an academician of the Ukrainian Academy of Sciences, a leading historian of early Soviet computers, and the man largely responsible for the MESM story being preserved. He, too, remembers the place very fondly: "Feofania was like heaven on earth. There was beautiful forest around the monastery with lots of oaks; there were beautiful lakes there and people who worked in the laboratories went to these lakes to have a swim during breaks, and the first to run to the lake was Lebedev! And a person could get a lot of mushrooms, strawberries in the field, and there were lots of hares, lots of rabbits in the forest. There were lots of wholly natural springs in Feofania and you can find them there even now."

Work started towards the end of 1948 with fewer than 10 people on the team, and considering its meager resources made remarkable progress. Zinovy Rabinovich says they were completely committed: "We lived there for two years until the prototype began to work. We worked 24 hours a day, so some people went to bed and other people woke up and went to work, so it took only two years to create a totally new technology. Lebedev had already designed and elaborated all the principles of the creation of the computer, so he made all the drafts by his own hand and these principles were almost the same as the principles of von Neumann, though they were not known in the Soviet Union at the time." While that claim is questioned by some in the West, Boris Malinovsky firmly supports it, on the basis that the scant information published in Western journals at the time was inadequate to reveal or even imply "von Neumann architecture." There were also significant differences of detail in the way the Soviet version was implemented. Doron Swade studied this question in detail and concludes that "in terms of its logic and architecture MESM is distinctive. It is very likely that it was based on what was in the public domain about the ENIAC, UNIVAC, EDSAC, and so on, but if it was influenced at all the influence was minimal and based only on the very general knowledge that was in the air at the time. So the Russians' fierce defensiveness of MESM's originality is to a large extent well founded. In performance terms it was very similar to the West's machines; they were all fighting the limits of technology. So was MESM original? Almost completely, yes. Was its performance comparable? Certainly."

Be that as it may, MESM undoubtedly had the classic central controller, main memory (and a secondary "passive" memory, a large drum with magnetic elements), arithmetic processor, input, and output. When complete it contained around 6,000 valves and was 8–10 meters long and about 2 meters in height—they had to use a ladder to reach the upper valves. It was kept so secret that few people even in Kyiv knew of its existence.

One of the early problems was an affliction suffered by all the other early computers, the large amount of heat generated. Rostislav Cherniak, who worked on the control of the processing steps that linked the various components of MESM, believes the power consumption was 7 kilowatts, much less than the ENIAC but still a formidable amount (and similar to the Rand 409). "It was very hot in the computer room, so we needed some cooling. We demolished one wall to make the room bigger, but that wasn't enough, so then we took the roof off as well." It's unclear from his account but it seems he meant the ceiling rather than the outer roof of the building, although he does say that working conditions were hard. That interpretation is supported by Boris Malinovsky, who came to Kyiv when the machine was already working. If he had missed the initial hard work of rebuilding the monastery-cum-laboratory, Malinovsky had certainly not been unscathed by the war, wounded twice while serving in the Red Army and losing his elder brother, a tank commander, in the Battle of Kursk, the greatest tank battle of all time. He says that "when MESM was switched on the temperature rose to more than 50 degrees centigrade [120 Fahrenheit]. The machine was situated on the ground floor and the house was two stories, so the ceiling was removed to make more ventilation."

However, these problems were eventually overcome and the machine began to work on November 6, 1950, the anniversary of the liberation of Kyiv. The team was able to send a brief report of their success to the Institute of Electrical Engineering the next day, the anniversary of the Revolution. This was quick progress from a 1948 start with a team that still numbered only 16.

The first full test program had a surprising result. Like some of its Western contemporaries, this was a ballistic problem. In order to check its accuracy, Lebedev engaged two renowned mathe-maticians to independently carry out the same calculations by hand. They had to work them out in binary code, step by step, so that they could compare directly with the results of each of MESM's individual operations. Malinovsky takes up the story: "They were seated in different rooms, and they couldn't speak to each other. When the machine began to calculate, everything coincided with the calculations made by hand, but then the machine began to give an answer that had a 1-plus, the answer plus one. So they began to check the work of the machine, but when all the tests were made everything was right. It was in the evening, so Lebedev told everybody to go to bed, sat down, and began to calculate this problem by himself. In the morning he came to the lab smiling, his glasses on his forehead, and he said the machine was right, and these two mathematicians who were sitting in two different rooms made the same mistake! So you can see from the very beginning the machine showed its advantages as compared to the man."

This was a considerable triumph. MESM was already much quicker than the mathematicians; now it had shown emphatically that it was more accurate as well. During 1951 it was put into full-time operation, and on December 25 that year it was formally accepted by the Soviet Academy of Sciences. There would still be a good deal of downtime for maintenance and further development, but this formal acceptance marked a symbolic step from a project to a usable machine.

Development continued, Rostislav Cherniak saying, "We moved step by step, we increased the memory, and we improved the quality of some of the peripheral devices. We had started with our own programs, then other people came to try out their problems. They came first from the Academy of Sciences and very soon some people came from places with big secrets and we were excluded from the laboratory." The "big secrets" were almost certainly atomic bomb calculations.

Boris Malinovsky says that MESM solved the most important problems of that time. "In 1952 it was the only working machine in the Soviet Union, so this computer solved problems concerning the design of the hydrogen bomb. It also solved problems connected with the calculation of rocket trajectories and satellite orbits and the design of electrical networks for long-distance power transmission. Besides this it was a machine that was used for research by other colleagues in the institute. And this machine worked up to 1957, and then it was brought to the Kyiv cinema studio, where it was used [as a prop] in the making of science-fiction and fantasy films."

Although the team worked hard, they also found time to enjoy themselves. Cherniak says, "We usually had a break for lunch and we used it to play volleyball. We would talk about the future of computers and Lebedev foresaw that one day a computer would not need a big building but could be housed in a small matchbox."

Viktor Ivanenko was one of the early users. As a graduate student he had calculations to run for his thesis and so he spent some time at Feofania, which he remembers well for other than technical reasons: "When I ran my problems here I had to do very complicated programming because the computer was very basic, with no printers at that time and special cards to program— punched cards. So I was helped by two girl programmers, and I remember how we swam in the lake and there was a very nice Russian sauna where you can use water and birch twigs to beat your body, so it was not a sauna in the European sense; it was a Siberian sauna. I was young. I was not married at that time. It was nice. It was a promising time!"

The girls were probably the two women assistants in Lebedev's original team, Dr. Ekaterina Shkabora and Ivetta Okulova. Just as in the Western projects, the pioneering teams were mostly, but not exclusively, made up of men, and programmers were often women. Dr. Shkabora was one of Lebedev's main assistants on MESM and she was later honored with the Sergei Lebedev Prize by the Ukrainian Academy of Sciences. Okulova was another of his first assistants and she went on to become a programmer on the SESM computer, the "Special Electronic Calculating Machine," which was conceived by Lebedev but developed and constructed at the Kyiv laboratory after he had left for Moscow.

Work continued on improving MESM after Lebedev's departure. Ivan Parkhomhenko joined the team in 1951 and his first job was to build something to print out the results. In its first incarnation MESM's results were displayed by flashing neon lights on the console. Parkhomhenko says, "You could read it like that but it was very difficult for the operator to retain this information and write it down fast enough. So I worked on that problem using a cash register like they used in shops to print receipts. The paper was in rolls and this was the first printer. We used electromagnetic "pushers" to operate it and this was quick enough because MESM was quite slow, about 5,000 bytes per second. It was a simple device but it worked!"

Other improvements were considered as well. As with other valve-based computers reliability was a problem, and Lebedev suggested to the young Malinovsky that he do the dissertation for his master's degree on one possible solution: "He suggested that I look at the use of ferrite cores instead of valves. I managed to do it, wrote the dissertation, and Lebedev was my examiner in 1953. It was approved." He went on to get a Ph.D. in 1962 and was elected to the academy in 1969. Ferrite cores were tiny iron rings with fine wires through them. They could be magnetized by passing a small electrical current through the wires in one direction to represent a "1" or magnetized in the opposite direction to represent a "0." They were subsequently used in a number of Soviet computers, not just for memory but as logical devices, in one design allowing the elimination of valves completely.

Just as in the West, there is no consensus over who made the first computer in the Soviet Union. One of the leading contenders was Isaak Bruk (or Isaac Brouk, depending on how his Cyrillic name is anglicized), who was born in the same year as Lebedev, and also took an early interest in the possibility of applying computing machines to advanced mathematics. At first he was thinking of mechanical computers, and before the war he participated in the design of a "big mechanical super-integrator," which was similar in principle to the differential analyzers made in America and Britain around the same time. However, in 1946, the Russian journal *Mathematical Tables and Other Aids to Computation* published its first brief descriptions of Western computers. Although short on technical detail, this prompted Bruk to consider electronics for computing and by that time he was already a noted expert on electrical power supply networks.

In 1947 Bruk got authorization to start work on his own design, the M-1 computer. Rather less is known about the circumstances of the construction of this than of MESM. He began work in the summer of 1948, producing a design that he patented in December the same year. Patent no. 10,475 was for "the invention of the digital computer" and Bruk's present-day advocates claim this as proof that he was the first, but it can't be regarded as conclusive. The patent was granted on a design only, not a working computer, and there is no evidence that Lebedev even considered applying for a patent or that he knew of Bruk's patent.

Like John Atanasoff on the ABC and John Mauchly with the ENIAC, Bruk teamed up with a younger engineer to turn his idea for a computer into reality. Bashir Rameyev was born in May 1918, just months after the Revolution. His grandfather Zachir had been a member of the pre-Soviet Parliament, a wealthy businessman and a famous Tartar poet. His father, Iskander, had become the chief engineer of the family gold mine and invented a completely automatic gold-processing factory, a one-man operation. None of this family history endeared the Rameyevs to the Communist Party and in the late 1930s Iskander was thrown into prison, where he died. That made Bashir a "son of an enemy of the people" and he in turn was thrown out of university. With that label it was difficult even to find a job, an extraordinary situation for a man who had been elected to the USSR Society of Inventors at the unprecedentedly early age of 17. That had followed one invention in particular, a radio-controlled armored model train that not only chugged along the model tracks but could fire its guns and deploy a smoke screen. It was his later inventions in the service of the war effort that rehabilitated him, though his subsequent career was still overshadowed and sometimes blocked by his father's history and his own technical weaknesses resulting from the unfinished university course. Fortunately his achievements saw him transferred from military service to a closed research institute headed by Axel Berg, who was more interested in bringing on talented younger men than in who their fathers were. When Rameyev, riskily listening to the BBC World Service in 1947, heard about the American ENIAC, he told Berg, who grasped the point at once. Berg recommended Rameyev to Bruk who took him on in May of the following year.

They worked on the project for little more than a year before Rameyev was suddenly drafted back into the army as a radar specialist (probably due to concern over the situation in Korea). After Rameyev had been languishing in the Far East for some months, Bruk managed to get him back to Moscow, but instead of rejoining him, Rameyev was made head of the laboratory at Special Design Bureau 245. He became in effect the chief designer of the

STRELA ("arrow") computer, officially led by Yuri Bazilevsky. This appointment required ministerial approval as Bashir Rameyev was still deemed "politically unreliable" because of his father.

Rameyev's departure undoubtedly hampered Bruk's progress, but he had a talent for spotting and developing bright young students. He had already assembled a team of eight (in addition to himself and Rameyev) from the Moscow Institute of Engineering. They had no previous knowledge of computing but learned quickly and several went on to specialize in the field. Ironically one of these was another "son of an enemy of the people," Nikolai Matyukhin, whose engineering brilliance meant he became effectively the chief designer, leaving Bruk with the role of scientific expert. At Matyukhin's suggestion a lot of the electronic valves were replaced by German war-surplus copper oxide diodes, a move that arguably makes the M-1 the world's first computer to use semiconductor logic, a decade before it became the norm. Much of the machine was built from surplus German military equipment, seized as reparations. They were also joined by Tamara Alexandridi, later to become Matyukhin's wife. Unusually among the female pioneers, she worked on the hardware rather than programming, designing the memory devices as part of her student diploma.

It is intriguing to note the parallel between Rameyev and Matyukhin's travails in the Soviet Union and the witch-hunt suffered by the Eckert and Mauchly team in America around the same time. On opposite sides of the world, ideological purity sharpened by Cold War paranoia labeled good patriotic scientists and engineers as politically unreliable and delayed some of the very work that was so important to both countries' economic and military future.

Isaak Bruk's enthusiastic young team, now without Rameyev and about the same size as Lebedev's crew, started building the M-1 in 1949 and worked quickly. On December 15, 1951, Bruk formally reported to the Academy of Sciences that it was finished, which would mean it was completed 10 days before MESM was declared fully operational. However, it's doubtful whether the M-1 was in a comparable state of readiness to MESM on that date. Certainly the M-1 needed three or four months' more work before it began its first major calculations, on nuclear reactor design. This illustrates the ultimate futility of asking who was first and expecting any kind of agreement on the answer (or even the question). As in other countries, both Bruk and Lebedev deserve great credit for their foresight and pioneering work, particularly as the Soviet authorities were yet to show any real enthusiasm for computing. They followed strikingly similar paths through life, born in the same year, graduating in the same year, developing their first computers in parallel, and dying in the same year.

Both were overshadowed for some years by a political decision to honor a third pioneer, though he has much less of a claim. In the words of the Russian computer historian Sergei Prokhorov, "Soviet history allowed only one victor." That was deemed by the authorities to be Yuri Bazilevsky, who, assisted by Rameyev, had produced the STRELA, which seems to have been the least impressive of the three pioneer efforts. It was used for at least one important task, the modeling of nuclear explosions, but more to the point it was the official project of the Ministry of Radio Industry. Despite getting preferential treatment, like supplies of scarce valves (Bruk having made do with German war surplus), the STRELA was markedly inferior in performance to both the M-1 and the MESM. Still the officially favored Bazilevsky was awarded the Stalin degree and honored as a hero of socialist trade. It would be some years before Sergei Lebedev was properly recognized by election to the Soviet Academy of Sciences.

Isaak Bruk was also up for the post of academician but lost out again. Part of Bruk's problem seems to have been that, although an outstanding designer who was very generous with his friends, he could be blunt with his superiors, never a good way to advance oneself in the Soviet Union. He was scornful when they were slow to embrace electronic computers, and he derided mechanical devices as "Stone Age." He applied the same label to the STRELA, with some justification, but that was just the sort of bluntness that didn't help his career. "The moral is you must not only be a good designer, you must not only be a good man, you must also be a good politician," says Prokhorov, adding, "In Russia, history is rewritten for the winner."

Bashir Rameyev left the STRELA project after the first one was completed, to head another computer center in the southern city of Pensa. There he designed the first of the URAL series, one of the first developed as a popular universal computer for industrial use, and a much more respected and successful machine than the STRELA. Bazilevsky continued working on his designs for several more years and produced seven STRELAs in all, though apparently doing little more to justify his early prominence.

None of the political maneuvering held Lebedev back. He had only ever conceived of MESM as a prototype, though it turned out to be a very useful tool for many years. His next step was the BESM, which stood for *Bystrodeystvuyushchaya Elektronnaya Schetnaya Mashina*, or "High-speed Electronic Calculating Machine." In 1951 he had been recalled to Moscow, to become head of the computer department of the Institute of Precision Mechanics and Computer Technology (and later its director). There he built BESM-1, which was completed in 1952, incorporating all the lessons learned while constructing MESM, and was a huge advance on the smaller machine. One statistic among many illustrates the progress made. MESM, partly because of economic limitations, worked at just 50 operations/second (ops). BESM-1 managed 1,000 ops in its first form, rising over the next few years to 8,000-plus. It was pretty much the equal of anything in the West at that time, though it's fair to say it wasn't designed in the same isolation as MESM. Rather, according to Malinovsky, it was built "only after a detailed study of the world's experience in designing super-high-performance computers."

Isaak Bruk was also hard at work in the early 1950s, and indeed computer projects sprouted all over the Soviet Union. Lebedev didn't stop at the BESM-1, instead producing a series of ever more advanced BESMs and a whole range of other special computers. One of the most impressive of the first generation (pre-transistor) was the M-20. With a processing speed of 20,000 ops, it was the fastest machine in the USSR and one of the fastest in all Europe. The Ukrainian effort continued with the KIEV computer in 1957, while Bashir Rameyev produced the URAL-2, 3, and 4 between 1959 and 1961. The first three in the Minsk series appeared between 1960 and 1962, and this series would become the most prolific in the USSR, some 4,000 being made.

One of the most unusual of Russia's 1950s computers was the SETUN, an astonishing design with no Western counterpart. It used ternary arithmetic, which meant that instead of the 1s and 0s of binary it used 1, 0, and −1. Although valves were the norm and now making way for transistors, the SETUN used ferrite cores, the same devices that Boris Malinovsky had looked at some years earlier. Again this started as an alternative to using fragile, power-hungry, and unreliable valves, and the original intention was to use binary arithmetic just as every other computer did. But in the course of its development the SETUN designer Nikolai Brusentsov realized that binary arithmetic was wasteful and that magnetic cores were better suited to the three states required by ternary logic. And ternary logic, he believed, was better suited to programming.

Brusentsov first devised and built a prototype that would add together two numbers using ternary arithmetic. His boss at the

Moscow State University's computer group, Sergei Sobolev, was so impressed that he promptly authorized the development of a full-scale computer. Brusentsov started the drawings in mid-1957 and assembly took place over the next year. Remarkably for the period, particularly considering the novelty of the design, it took only 10 days of testing and adjustment to become fully operational. It proved from the start to have low power consumption, and it was reliable and reasonably cheap to build and sell. About 50 SETUNs were manufactured and used widely, mainly in universities. A follow-up, the SETUN-70, was made in 1968, but by that time the freedom to pursue outlandish designs was coming to an end and in any case the Moscow State University decided to get out of computer production. Perhaps crucially, the first SETUN came into use just as valve-based computers were coming to an end, and the shift to transistors offered much the same advantages in terms of reliability and power consumption as ferrite cores. Ternary logic alone wasn't enough to make the SETUN into the world-beater it might have been, although it is said that when the first one was taken out of service in 1975 it had worked those 17 years without a single malfunction. Sadly it was then destroyed, as were nearly all the others, though Brusentsov managed to save a SETUN-70 by secreting it away in the attic of one of the university buildings.

By the mid-1960s Lebedev was still in the forefront of Soviet computing, and in 1966 his team completed the last of the BESM line. The BESM-6 was a supercomputer of the second generation (i.e., transistorized) that stood comparison with the leading Western models at the start of its life in 1967 and was to remain in production until 1984. Among its many important applications it was the heart of the AS-6 space flight control system used for the joint USA–USSR Apollo–Soyuz project in the early 1970s. Boris Malinovsky continued his work in Kyiv, maintaining the city's prominence with the DNEPR (named after the river running through the center of the city), which was the Soviet Union's first general-purpose control computer. In the second half of the 1960s came two important designs of minicomputer: the NAIRI from the Institute of Mathematical Machines in Armenia and the MIR from Viktor Glushkov's Institute of Cybernetics in Kyiv.

This whole rich, diverse effort was genuinely independent of Western computing. But for the intensity of the Cold War, there would undoubtedly have been some cross-fertilization, in both directions, of techniques and technology. But the Soviet Union's computer development was not only largely independent of the West, it was also composed of a collection of jealously independent teams. By the 1960s the Russians had an array of incompatible computer designs, mostly produced in small quantities. Software written for one had to be rewritten to run on another, along with retraining of both production and maintenance engineers.

By now the so-called "third-generation" computers were appearing, using integrated circuits in place of many individual transistors. This was an appropriate time to tackle the widespread incompatibility of hardware and software that was limiting the application of computing. Again a political decision was made, to create a new project of compatible computers that would be called the ES, or Unified Series. In January 1967 Soviet officials formally suggested copying the logical structure and command system of the IBM 360. This would give Soviet computers a common hardware framework that would facilitate their mass production, deployment, and maintenance along with common software requirements (and access to much Western software). The downside was that neither hardware nor software was available directly, nor the technical detail necessary to make copies, because of US limitations on "technology transfer" to the Eastern bloc. The arguments over the decision to "make or take," as it was characterized, raged for three years, but at the beginning of 1970 the political choice was made to go for IBM compatibility.

It was a decision that had some logic to it, but it brought to an end a lot of innovation in Soviet computing and it was a terrible blow for Lebedev, now in his late sixties and already in poor health. It still rankles with pioneers like Boris Malinovsky, who says, "It was a great mistake. The Lebedev Institute and IBM were at about the same level of development at that time, but when the Soviet Union started copying previous models of the American machines its technology began to lag behind." Like many former Soviet engineers, Malinovsky believes the government should have authorized Bashir Rameyev's proposal for a family of URAL computers that would have paralleled the IBM 360 series. Instead only three versions of the URAL were commissioned and Rameyev quit his job as chief designer in disgust.

Looking back at the decision, Zinovy Rabinovich is rather more even-handed: "There were a lot of discussions about this here. Lebedev was absolutely against this copying, but some good came out of it. A lot of machines were produced. Maybe they were not as good as if we had designed them ourselves, but there were a lot of them and they satisfied the demands of much of our economy. But I think it would have been better to develop in both directions, one of borrowing, the other of original design."

This is another illustration of the remarkable global dominance that IBM had achieved by the end of the 1960s, despite its early reluctance even to consider the mass production of electronic computers.

It is easy for those who grew up during the Cold War to accept the stereotype of the cold, hard Russian—particularly those involved in science, technology, and the arms race. But it's clear that Sergei Lebedev was anything but cold; indeed he is still adored by those who survive. Rostislav Cherniak, when asked to say something about Lebedev in the garden of his dacha in the summer of 2001, opened his mouth to speak and dissolved in tears. Only after a break for one of the ripe peaches growing overhead and a cup of good strong Brooke Bond tea did he recover his composure enough to get the words out. "Lebedev was outstanding not only as a scientist, but as a person. He never raised his voice; he could not be nasty to anyone. It was very touching and we appreciated this because Lebedev worked day and night."

His last great project was the ELBRUS fourth-generation supercomputer, later tagged "Lebedev's scientific legacy" (the fourth generation was the next step after integrated circuits, when manufacturers started putting many thousands of components onto a single "chip"). Named after the USSR's highest mountain, which he had climbed many years earlier, this would take Soviet computing into the era of over 10 million ops (ELBRUS-1) and then to 125 million ops (ELBRUS-2). Lebedev created both projects but didn't live to see them in operation. In 1974 he developed pneumonia, not helped by general fatigue brought on by decades of working all hours, a workload he had scarcely reduced even when he turned 70. He was also demoralized by the decision to adopt the IBM 360 standard, combined with official resistance to his efforts at European collaboration (West as well as East). Pneumonia was too much for his weakened constitution and he died on July 3, 1974.

His memory lives on in those who regret that the Soviet Union was so secretive about its achievements in computing for so many years. Zinovy Rabinovich says quite simply, "Lebedev was an absolutely unique person. I never met anyone like him in my life. He stands absolutely apart in his intellectual abilities." Boris Malinovsky agrees, saying, "Lebedev was an extraordinary person and nobody can be compared to him." And Doron Swade says quite simply, "What emerges is Lebedev as the towering figure in

Russian computing and there are several reasons for that. He was absolutely brilliant, he was politically quite astute, and he is also a figure who almost uniquely straddles and dominates several generations of the most populous and influential machines."

CHAPTER 7

WIZARDS OF OZ

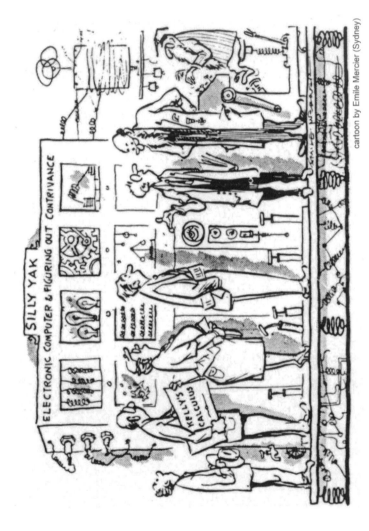

"They want to work out the figures that determine the winning number in a jackpot lottery"

The comic potential of Australia's second computer, the SILLIAC, was spotted at once and exploited by, among others, the great *Sun* cartoonist Emile Mercier (by permission of Mrs. Pat Mercier).

If the history of Soviet computing has been largely neglected in the West, there is another country whose computer pioneers have been even more comprehensively ignored—Australia. Yet this is a country that boasts, among many more technical achievements, the only computer pioneer to have a crocodile named after him. One of the world's first few real computers (electronic, digital, stored-program) ran there for the first time in 1949 and the Australians made it themselves. It didn't come out of the blue of course. Australia, like all the countries in early computing, had a history of ingenious mechanical calculating devices. For instance, it is the home of the Julius Totalisator—a machine that worked out the bets at race courses. This is after all the nation that by its own admission stops for a horse race.

The man behind this invention was George Julius, born in England in 1873 but taken to Australia at an early age by his parents. As a boy he showed an aptitude for things mechanical that he later attributed to a grandfather who, he said, "was one of the court physicians in London, but had such a mechanical bent that he spent what money he had in backing any invention that had wheels on it." Little surprise then that Julius became an engineer when he grew up and set up Australia's first engineering consultancy.

In his 40th year, in 1913, he unveiled an invention that would change the face of on-course betting. This was an amazing contraption of wires, pulleys, and metal boxes that simultaneously issued betting tickets and kept a running total of the money placed on each horse. He had designed it first as an automatic voting machine, but the government wasn't interested. Then a friend told him about racing's need for an efficient means of totalizing bets, so he redesigned his vote-counting machine to become the first *totalisator* in the world, and four years later he formed the company Automatic Totalisators Ltd. He added electrical power to the machine and a decade later came the huge advance of automatic odds calculation. Now his totalisators would issue the tickets, tot up the money placed on each runner, and calculate the odds in proportion to the bets laid. By this time Julius's machines were being installed all over the world and George Julius had become the first chairman of Australia's Council for Scientific and Industrial Research (CSIR), which would later lead the country into the computer age.

Julius served on the Central Inventions Board during World War II and received ample evidence of his adopted countrymen's fertile minds, citing as just one example that "someone suggested we should freeze the clouds so that no bombs would drop through. He did not say how it was to be done." The remarkable Sir George Julius would not play any part in the electronic computer revolution, as he was already in his late sixties by the time the war ended, and he died in 1947. However, the company he founded, Automatic Totalisators Ltd., would almost 20 years later develop the world's first computer totalisator system and sell the first one to the New York Racing Association. And it was CSIR, the council that he had chaired for its first 20 years of life, that was the vehicle for Australia's first electronic computer project.

The inspiration for the project came from Trevor Pearcey, "the very epitome of a *Boy's Own Annual* boffin," with his "fuzz of unruly hair, broad brow, glasses, and detached demeanor," according to

Dr. Clive Coogan, then a student at the University of Sydney, where the computer was built. Like George Julius, Pearcey had been born in the UK, but little is known of his life before he graduated with a first in mathematics and physics from London's Imperial College of Science and Technology. He started a Ph.D. but gave it up because of the war and went to work for the Ministry of Supply, in the secret and important Telecommunications Research Establishment (TRE). There he applied his knowledge of advanced mathematics to Britain's wartime radar developments and became aware of the pressing need for better ways of carrying out large computations.

Pearcey also got some hands-on experience with the rare differential analyzers, like the machine that the Moore School was using to calculate firing tables in America (and prompting John Mauchly to conceive the ENIAC). One of these was at Manchester University, where he worked with Douglas Hartree, an important figure in early British computing. He used another differential analyzer at Cambridge, where he also became familiar with the Mallock machine, another kind of mechanical calculator used to solve linear equations. He worked with an Aussie expatriate, Leslie Comrie, whose own company in London, the Scientific Computing Service, made extensive use of mechanical punched-card tabulating machines and the like. For Pearcey it was all an excellent grounding in the current state of mechanical calculation and the emerging technologies.

Trevor Pearcey emigrated to Australia in 1945, though his passage took several months and involved a long diversion via America, where he saw the Harvard Mark 1 in operation. Fifty years later he recalled that visit from his hospital bed, saying he hadn't been particularly impressed with the Mark 1, because it was only a mechanical calculator. He saw that it could do about two arithmetical operations a second, which "was obviously not going to be adequate." He could see they would have to go electronic, which would immediately give a thousandfold gain in processing speed. During that 1945 visit he also realized the problems with mechanical input and output of data and wondered how those processes might be speeded up enough to allow the electronics to do the arithmetic at maximum speed. The electronic design, he said, wasn't such a challenge, as "we knew about counters and such things from radar. We knew about using vacuum tubes as switches." (Australians used the American term "tubes," despite their exposure to British "valves.") Just as in other countries, these techniques were part of the general knowledge of leading practitioners in the emerging field of electronics.

While passing through America, Pearcey also visited the Massachusetts Institute of Technology to see the Bush Differential Analyzer in operation. What he missed, though, was the unveiling of the ENIAC and the Moore School lectures, both of which happened the following year. He also knew nothing of the British code-breaking machines. Although he had done highly secret war work in the UK, even the existence of Colossus was very effectively restricted to those who "needed to know." By the beginning of 1946 Pearcey was settled in Sydney, Australia, where he started formulating his ideas for high-speed computing in among his other work at CSIR. As in other countries, World War II had provided a boost to technological development in Australia, with the need for radar and similar high-tech systems obliging CSIR to establish a Division of Radiophysics at the University of Sydney in 1939. With the knowledge gained both from these wartime activities and from other staff returning from service in the UK, the division was well placed to start looking at electronic computation as soon as the war ended. A major objective of CSIR was the conduct of research to assist the country's primary industries, and statistical techniques were an important part of this.

Pearcey produced a prodigious quantity of academic papers in his working life, on a wide range of subjects, including radio-wave propagation, optical physics, air-traffic control, crystallography, and many others. Early the following year he finally persuaded Edward Bowen (commonly known as "Taffy"), chief of the radiophysics division, that they should get into high-speed computing.

As well as starting work on what would become the CSIR Mark 1, he also established at Sydney the first university course anywhere in the world on numerical mathematics and methods for students other than statisticians. As elsewhere, there were no training courses for computer engineers, but this was in effect an introduction to computer theory, practice, and programming principles. Pearcey himself delivered the course every year from 1947 to 1952, and many of those who attended went on to make their careers in computing.

Pearcey was already thinking beyond the immediate needs of the scientific community. In 1948 he wrote that "in the non-mathematical field there is wide scope for the use of [computing] techniques in such things as filing systems. It is not inconceivable that an automatic encyclopedia service, operated through the national teleprinter or telephone system, will one day exist." Given that only a handful of people in the world at that time had foreseen any applications outside academia and the military, this vision was remarkable.

Once again there was a pair of pioneers at the center of the team. Pearcey's right-hand man was Maston Beard, an electrical engineering graduate of the University of Sydney. Together they defined the design by the end of 1947, Beard working on the hardware and Pearcey the theory. Another significant figure in this early effort was David Myers, who in the mid-1930s had also spent time with Douglas Hartree at Manchester University, where he, too, got the opportunity to use the differential analyzer. Even before the war there had been a great deal of interest among Australian scientists in more advanced calculating machines, and the knowledge that their peripatetic colleagues brought back from their foreign travels was much prized. On his return to Sydney, Myers became chief of the Division of Electrotechnology at CSIR. Myers proposed a specialist section to meet the need for high-speed computing and in 1948 he became head of the new SMI, the Section of Mathematical Instruments.

At around the same time, the Australian engineers first became aware of the parallel developments in Britain—Wilkes's EDSAC in Cambridge, Williams and Kilburn's Baby in Manchester, and Turing's ACE at the National Physical Laboratory in Teddington. Pearcey returned to the UK at the end of that year to look at all three but, following his visit, he decided to leave his team's design unaltered. Writing in 1988, he asserted that the CSIR Mark 1 "was completely 'home-grown' some 10,000 miles distant from the mainstream development in the UK and USA." It seems the parallel developments were defined by their similar starting points, similar needs, and the limits of available technology. It was in a sense Darwinian evolution applied to technology—"natural selection" brought about similar solutions.

CSIR also sent David Myers to see the UK projects towards the end of 1948 and he returned with the conclusion that all the currently available storage methods—mercury acoustic delay-lines, rotating magnetic drums, and the Williams–Kilburn tube as used in the Manchester Baby—were unsatisfactory. No doubt he would also have heard of the American problems with Selectrons and iconoscopes. Myers's response was to recommend that computer development should be restricted to components only, until some more effective method of storage emerged. This undoubtedly held back the Australian effort for some time, though they wisely chose to continue developing the most practical solution at the time, mercury delay-lines as used in Australian wartime radar (just as they were used in Britain).

Despite Myers's reservations, Pearcey's team made progress. Peter Thorne, who worked on the computer in later years and co-wrote its definitive history, *The Last of the First*, says that this was due in no small way to Reg Ryan, the man who had the task of turning the mercury delay-lines into a working computer memory. This was done quite independently of the similar development by Tommy Gold in Maurice Wilkes's team at Cambridge in the UK. Reg Ryan was, says Thorne, "a young engineer, younger than the others, and I suspect they did to him what people still do today. They didn't tell him what he was doing was possibly fairly tricky; they just told him they expected him to do it! He designed the whole memory system and that in itself I think was pretty remarkable." Ryan's biggest problem was that the acoustic pulses, which represented the data, traveled round the delay-line at different speeds as the temperature rose and fell. The solution he devised was a dummy line in the mercury tank, using measurements from that to control the clock speed of the computer in a feedback loop. That way he could keep the computer's processing unit synchronized with the data.

By 1949 they had a working machine. The date of the first run of an early computer is often hard to decide with any certainty, owing to the problem of definition, and it is particularly hard to decide when the CSIR Mark 1 first operated. But there is little doubt that towards the end of 1949 Pearcey wrote some basic programs and Beard used them to run a sequence of mathematical operations, probably in November (the 24th is given in some sources as the actual date, but the evidence for that particular day is unclear). It was a pretty basic machine at that point but it joined the British Baby, the EDSAC, and the American BINAC in the first quartet of genuine stored-program electronic computers. By that time it had become the CSIRO Mark 1, after the parent body was renamed the Commonwealth Scientific and Industrial Research Organisation.

Development continued and in 1951 the CSIRO Mark 1 claimed a true first when it was programmed to play some music. This was no gimmick based on a few notes of dubious musical quality, but a genuinely imaginative piece of complex programming. There were no sound cards, nor even digital to analog converters. The programming team was Trevor Pearcey and Geoff Hill, who was from a very musical family and whom Peter Thorne calls "Australia's first great computer programmer." They had to create raw data pulses, which were sent to an audio amplifier connected to a loudspeaker, but this was an advantage as the programming was at such a basic level that almost anything was possible in terms of timing, frequency, and so on. In August 1951 the CSIRO Mark 1 was demonstrated at the country's first Conference of Automatic Computing Machines, at the University of Sydney, though it was still incomplete. It was one of the highlights of the conference to hear the computer playing popular melodies like "Colonel Bogey" and "Bonnie Banks." It was a startling demonstration of the potential of computing to spread into fields beyond the scientific and military arenas. In the opinion of one of today's leading experts on the subject, Paul Doornbusch of the Sonology Institute at The Hague in the Netherlands, this was the first proper computer-generated music in the world.

An honored guest at that first conference was none other than the pre-eminent mathematician Douglas Hartree. Like Trevor Pearcey, Hartree was an expert in the application of mathematics to theoretical physics. (Pearcey later recalled his former mentor passing the time on long train journeys by solving complex differential equations using only a slide rule.) During his visit to Australia, Hartree was asked by CSIRO to advise them on the future development of computing and he told them to set up a

Division of Applied Mathematics. There was really only one person with the abilities and stature to head such a body and that was John Jaeger, then professor of mathematics at the University of Tasmania. However, Jaeger was more interested in geophysical research and declined the offer, and the plan lapsed. "Thus passed the first opportunity for establishing effective research and development and a possible future industry for computing in Australia," concluded Pearcey sadly.

More missed opportunities were to follow, but in the meantime development continued and by 1953 the CSIRO Mark 1 was in almost constant use. It used components similar to those in other parts of the world, though largely developed and manufactured in Australia. The memory was a bank of 32 mercury-filled acoustic delay-lines, with auxiliary storage in the form of a rotating magnetic drum. At first, data were entered and results output by punched-card machines and teleprinters, for compatibility with existing equipment, though before long the CSIRO moved to faster paper-tape systems. To begin with, the Mark 1 used around 2,000 tubes from the country's own radio industry. Later many of these tubes were replaced by the new semiconductor diodes that were just coming into production in the early 1950s. Even so, heat dissipation was as usual a problem, with the machine consuming 30 kilowatts, and cool air had to be blown from the basement through the computer cabinets and into the outside world.

Other problems proved more challenging than just keeping cool. One was the appearance of random digits in the acoustic-delay memory. The researchers were helped in tracing this one by realizing that when the random digits appeared they were spaced three seconds apart. After some head scratching they found that a meteorological radar mounted nearby was rotating at one-third of a turn per second. Each time its signal passed the air-conditioning duct on the roof, some of it was reflected down into the guts of the machine. A suitably high-tech solution was found—the mouth of the duct was covered with fine-mesh chicken wire that appeared impenetrable to radar of that wavelength, but didn't impede the flow of air.

However, solving such problems didn't guarantee a future for the CSIRO Mark 1 project. When development finished in October 1952, the team invited four Australian electronics firms to tender for the manufacture of commercial versions (just as Ferranti was already commercializing the Manchester Mark 1/MADM). Only two companies responded, and nothing came of the exercise. In 1953 and 1954 the CSIRO Mark 1 was in full-time operation in Sydney, but other computer projects were initiated in those years and the Mark 1's design began to look out of date.

Worse still, the CSIRO management was losing interest in computer development, choosing instead to concentrate on its radioastronomy, in which it was a world leader, and cloud physics, because of the potential economic benefits of rain making. It also still had research into primary and secondary industries (such as agriculture, forestry, building materials, and coal) as its major objective, in spite of the rapid post-war growth of more advanced manufacturing industry and the attendant need for sophisticated computing. Even though the radiophysics divisional head, Edward "Taffy" Bowen, had been persuaded at the start of the project that CSIRO should be in the latter field as well, he still saw it only as a tool that would assist the main areas of research, not as something worthy of research in its own right.

Doug McCann, a science historian at the University of Melbourne and co-writer (with Peter Thorne) of the CSIRO Mark 1 history, says this was certainly not the attitude of Trevor Pearcey, who was "acutely aware of the promise and potential of computing as a discipline, a technology, and an industry in its own right." Pearcey believed he just couldn't get through the indifference shown by the radiophysics division management, even though that successful first run in 1949 had put Australia at the forefront of computing developments along with the US and the UK.

However, Peter Thorne is not so ready to accept the "flip explanation" that CSIRO simply preferred research into rain making rather than computing. "It's a good story but I think there were two other factors involved. One was Pearcey himself. I worked with Trevor and knew him quite well, a very clever fellow but very much the rather reserved scientist, and he could be an awkward person. He was one of those people who would walk through your lab, where he didn't need to be, look at something you were doing, make a critical comment that would annoy the hell out of you, and walk on, and about three months later you'd realize he was right! He was just one of those people with a quick mind, great insight, and quite charming but not always tactful, very much the scientist of that period. He certainly wasn't a public relations man and I think the project he had started needed more PR push to get support and it probably needed a bigger team. It needed selling skills that Pearcey wouldn't have claimed to possess and I don't think he would even have admired them."

The other significant factor, argues Thorne, was the arrival in 1953 of a Canadian physics professor at the University of Sydney, where the radiophysics division of CSIRO was based. This was Harry Messel. According to Thorne, he was "a hard-drinking, hard-driving sort of guy, a very dominant figure, and the echoes of his influence reverberate round the nation even today." Messel displayed astonishing energy throughout his life and does so well into his retirement. Born of Ukrainian parents in Canada in 1922, he says he was "brought up with a very healthy lifestyle in the prairies, hunting, trapping, and fishing." He served as a paratrooper in World War II, marched through London on VE Day, and promptly volunteered for the Canadian–American task force that was due to parachute into Tokyo the following New Year's Day. The atomic bombings put an end to that plan, so "I got out early in 1946 and went straight to Queen's University in Kingston, Ontario, and enrolled in two degrees simultaneously, one during the day, one at night. I think I was the only person they ever allowed to do that and I graduated with honors in engineering physics and in mathematics."

Postgraduate studies followed, first at St. Andrews University in Scotland, then at the Dublin Institute for Advanced Studies in Ireland, where he gained a Ph.D. under the great theoretical physicist Erwin Schrödinger. Then in 1951: "I decided I'd come over to Australia to take a little look," accepting a senior lectureship at Adelaide University which lasted only eight months. "I had a fall-out with the vice-chancellor. I had made a proposal we set up an institute for advanced studies and bring Australian students back to Australia, so they get their Ph.D.s here rather than go over to England and America—and, of course, they keep the best ones over there and send back the ones who are not quite up to scratch. The premier of South Australia thought it was a great idea and I made a speech in the town hall and then the next morning the vice-chancellor came up to me and said we had enough of sciences; what we need is more of the arts."

Messel resigned within three days. "I wasn't going to muck around because I had the world at my feet. I was going back to Europe but on the way I had to pass through Sydney and I was met at the airport by two distinguished professors, saying that the vice-chancellor of Sydney, Stephen Roberts, would like to meet me. So I met him, he offered me the chairmanship of the school of physics, and I made a preposterous set of demands, which I don't think anybody has ever equaled: that I be allowed to make 14 permanent academic staff appointments, that I be able to get the journal *Physical Review* and all the British physics journals flown in by air, and so on." Believing that they wouldn't give all that to a brash young man still in his twenties, Messel flew back to Europe, but "I got to Milan in Italy and there was a telegram saying, "The University of Sydney has accepted all of your recommendations and you are hereby appointed the Head of the School of Physics with effect 1st Sept. 1952." So there it was. They called my bluff!"

Messel was particularly interested in nuclear physics and promptly set about establishing a department in the subject at the university. That in turn demanded advanced computing and it soon became obvious to him that even a new and improved version of the CSIRO Mark 1 would not be adequate. "Taffy Bowen [head of radiophysics] and I got on together rather well and he told me the Mark 1 wasn't bloody working at all, it was a pain in the ass, they couldn't get anything from it, it was draining their money, and Taffy wanted the money for his cloud-seeding experiments. So he said he would do everything in his power to help me get a computer; then he could get rid of the bloody CSIRO Mark 1! And that's exactly what happened. It was just a very difficult machine, you know, and it never did produce anything of any consequence at CSIRO."

Another reason for rejecting an advanced version of Pearcey's CSIRO Mark 1 was the still-unresolved problem of data storage, with neither the mercury delay-line nor the rotating magnetic drum allowing sufficiently rapid access to data. Manchester's Williams–Kilburn tube was much quicker and growing in reliability. It had been adopted in several US projects, most of them similar in principle to von Neumann's Institute for Advanced Study machine, the IAS. One of those projects was the ILLIAC, so called because it was built at the University of Illionois. So Harry Messel set about applying his considerable PR talents to getting the support for a new machine that would be based on the ILLIAC. This gave Bowen the excuse he needed.

Thorne says now that "it was clear that Messel had got a fair bit of momentum and was going to go ahead and build this parallel machine [the CSIRO Mark 1 used serial processing, which was slower than parallel]. It was going to be a lot faster, based on more recent technology, and it was going to be next door as it were—on the same campus. That probably meant that if you were the head of CSIRO radiophysics and you had to decide where your funding priorities were, you wouldn't put your money into supporting what was already a somewhat aged computer."

Thus, early in 1954 the radiophysics chief, Edward "Taffy" Bowen, decided to bring the CSIRO Mark 1 computer project to an end and it was officially terminated on April 13. After considering a number of options, the institute decided to donate it free to the University of Melbourne, which had no computing service. It was a generous offer to a rival city, and university, as it was valued at £75,000 (at a time when £1,000 was a good annual salary).

Dr. Frank Hirst was already in place at Melbourne and he was asked to collect it. He didn't know much about the machine at first, but when he realized just what it was capable of he couldn't believe they were letting it go—"I thought to myself CSIRO must be mad giving this away!" He also recalled that Pearcey seemed very unhappy it was going, but still told him "anything you want in this room you should take with you." Hirst was regarded by his colleagues as a very resourceful man, even into his retirement, and he took everything in that room, even the blinds and light fittings (40 years later they were still turning up in the computer department at Melbourne, often in their original cardboard boxes). In the middle of 1955 the whole lot was loaded onto a large truck and driven down to Melbourne. There it acquired another new name. "CSIRO Mark 1" had probably owed something to the title of the "Harvard Mark 1" that Pearcey had seen some years earlier, but it didn't exactly trip off the tongue. Almost every computer in the world by that time ended in "-AC" so it duly became CSIRAC (pronounced "SIGH-rack").

Reinstallation in its new home in Melbourne was a major task and it was many months before CSIRAC ran again. The official installation ceremony didn't take place for a full year, when the sparkling new Computation Laboratory was opened on June 14, 1956. In the meantime Frank Hirst had taken the opportunity to make a number of detailed improvements. He had been surprised to find Sydney treating the CSIRAC as if it was still a development prototype, with cabinet doors left open and untidy wiring. In Melbourne, Hirst and his colleagues made it look like a finished computer and it was soon pressed into full-time operation, carrying out some 30,000 hours of computation on around 700 projects over the next eight years. Only about 10 percent of its running time was taken up with maintenance, a good figure for the time. Not that it was particularly reliable and Peter Thorne reckons it "had a mean time between failures of probably about an hour," partly because the power system didn't have much reserve capacity. Just plugging in a kettle in the adjacent tearoom could trip the power circuit and shut down the computer. Another thing they never really conquered was the weather. Just as in Sydney, CSIRAC was mounted on a floor with a decent-sized basement beneath it and this was used as a source of cold air sucked up through the computer cabinets and blown out through the roof. But when temperatures hit the high 30s (centigrade) in the Aussie summer, it simply couldn't be adequately cooled and had to be switched off.

Frank Hirst went to some lengths to turn the recommissioning ceremony into a memorable event. It was no simple cutting of a symbolic ribbon. On the big day the chairman of CSIRO, Sir Ian Clunies-Ross, formally asked the vice-chancellor of Melbourne University, Professor George Paton, to accept CSIRAC and then pressed a button on the control panel. This switched off an electromagnet that had been invisibly holding a parchment scroll flat on the control panel, allowing it to curl up magically before their eyes. Professor Paton accepted the gift and in turn pressed his button. CSIRAC burst into life, printing out a message onto the parchment, which read:

> Mr. Vice Chancellor
> Thank you for declaring me open. I can add, subtract and multiply; solve linear and differential equations; play a mediocre game of chess and also some music.

While most of the early computer pioneers were, as previously noted, dismissive of attempts to anthropomorphize their machines, here is another example of the scientists themselves treating their computers as if they had not just human-like intelligence but even a personality. Hirst was very relieved when the message printed out faultlessly. It might look like the simplest of tasks today, but in 1956 with a program held in rather unreliable memory that would still spontaneously alter itself, fingers were firmly crossed as the ceremony was played out in front of the most senior men from the two institutions. Eight years later Hirst would be the one to switch CSIRAC off for the last time at another, rather more poignant, ceremony.

Back in Sydney, Harry Messel's project to build an Australian copy of the American ILLIAC was well under way. First he had to get the design of the computer from the University of Illinois and permission to copy it, so he did a deal with them. "I had very good connections with the people in America, especially the Atomic Energy Commission, and they were supporting the upgrade of ILLIAC, the Illinois computer that was based on MANIAC at Los Alamos. So I approached Illinois University and the AEC to see whether we could get the plans and they said, 'Well, if you send over staff and help with design of the next generation, you can have the blueprint.' Sure enough that's what we did: I sent over two staff to help with the redesign and in 1954 they came back to Australia with the very latest in computing."

Also involved with the ILLIAC in Illinois was John Blatt, whose Austrian-Jewish family had fled Nazi Germany for the USA in 1938. Blatt wasn't much happier in McCarthy-era America and he emigrated again, this time to Australia, where he arrived at the University of Sydney just as Harry Messel was reviving the institution's interest in computing. By this time Blatt was a renowned nuclear physicist with a major publication to his name, so his support was valuable to Messel. "I had to convince my own staff we needed a computer. Strangely enough the people who thought the computer was worthless were the radioastronomers and who are the biggest users of computing now? Radioastronomers! I love it, I love it!" If that thought still reduces Messel to helpless laughter, the name of the computer too was a source of some amusement. Sydney's ILLIAC was almost inevitably named "SILLIAC" and students quickly cottoned on to the humorous potential of the name. Even the famous *Sun* cartoonist Emile Mercier once took the "Silly Yak" as his subject.

The machine itself was no joke. Although based on the ILLIAC, it was again built entirely locally and in less than two years. Funding came from a friend of John Blatt's, a racehorse owner and jeweler, Dr. Adolph Basser. Again Messel's approach was straightforward: "Nobody in the university was going to give me money for the computer, so I made contact with Basser and told him about it over lunch. He was racing at the Melbourne Cup in 1954 and he won £50,000 and he decided to donate it to my school for the construction of the computer. So that was how it started and the next year, when we were running a bit short, he won the Melbourne Cup again and he donated that same amount again! It paid for the whole thing; it was wonderful." Assembly finished on June 13, 1956, and the first program ran just 11 days later, showing one of the advantages in using a proven design, though Sydney made some modifications to suit their specific requirements.

International cross-fertilization had continued during the project with the earlier arrival of Barry de Ferranti, a Queensland University graduate who had spent several years as Maurice Wilkes's first research student on the EDSAC. De Ferranti followed that up by joining the British company that coincidentally bore his name and he played a major role in defining the coding and programming of the Ferranti Mark 1 in Manchester before returning to Australia.

The SILLIAC was given its own home in the newly created Adolph Basser Computing Laboratory (the name being due recognition for its major benefactor). Amazingly even now, but particularly for the time, the computer was complete and operational six months ahead of schedule. It gave excellent service for 12 years, earning a formal decommissioning ceremony by the university vice-chancellor when it was finally time to switch it off in May 1968. For the last couple of years of its working life, it had been largely supplanted by an English Electric KDF9—the SILLIAC being reduced to the ignominious role of spooling the KDF9's results to local printers.

While the SILLIAC was put into regular academic service with great speed, it was narrowly beaten by the CSIRAC at its new home in Melbourne. CSIRAC's new owners treated it quite differently from the way it had been viewed in Sydney. There it had been a development machine, but the Melbourne team under Frank Hirst tidied it up and treated it as a workhorse, albeit one they revered. Among the many people who looked after CSIRAC during its 15-year life were several women. In 1959 Kay Sullivan, needing a job in order to work her way through a part-time degree, saw an ad for a technical assistant to work on the CSIRAC. "I didn't know what CSIRAC was but Peter Thorne (whom I had met at school) said, 'Oh, that's the computer—that would be great!' He had seen CSIRAC on a school visit. Anyway, I applied and was interviewed and then offered the job all in time to start lectures only a week late. I was told later that they had a large number of applicants and Frank Hirst came straight out after my interview and told the others, 'We'll have the redhead!'"

That informal attitude was, says Kay, "typical of the way we all worked. There were no secrets, no one ever closed their door, and all conversations and telephone calls could be heard by all of us. I don't remember any breaches of confidence into the world outside. I think we were all naturally discreet! It was a very inclusive atmosphere in which to work and quite social in the way good labs are. We all had morning and afternoon tea together at one table in the student café with no demarcation between Trevor Pearcey, Frank (who was a senior lecturer in physics at that time), the engineers (Ron and George), and the TA (me). They joined in the general discussions, anything from collating manuals to some deep philosophical debates on the future of computing. We had a healthy respect for each other's intellect and made a great team with loyalties and bonds which have lasted."

Unlike most of her female colleagues around the world, she wasn't specifically a programmer, as the TA's job was to "do anything and everything." One of her main roles, like that of everyone in the team, was to be the interface between the user and the machine, because CSIRAC wasn't very user-friendly and few users had any prior knowledge of computing. "There was a sense of mission in helping people to see some of the potential of computers and to understand that there was logic not magic at work."

It was shortly after Kay Sullivan joined the team that Peter Thorne got involved with CSIRAC too. He had been born in London in 1940, but his family had lost everything in the Blitz and emigrated to Australia in 1948. "My interest since the age of about 13 had been ham radio and electronics, so I hung around the computation laboratory. Frank Hirst thought I was a likely lad and suggested I become the weekend service engineer on CSIRAC—it needed warming up and possibly some repair, because it wasn't so reliable with 2,000 vacuum tubes. I was still completing my B.Sc. in physics, so that became my weekend job and it's been part of my life in one way or another since the early 1960s." So too has Kay Sullivan, as she became Kay Thorne during their time on CSIRAC.

The Melbourne team was very proud of their computer and was eager to show it off to the public. Demonstrations of its powers were frequent and it was a major attraction at university open days. Naturally the computer demonstrated its music-making ability and there were computer games of a very basic but addictive type, such as the infuriating "Nim" (see Appendix C). There was reaction testing, a "what day of the week was I born on?" calculator, and so on. Sections of punched paper-tape were popular souvenirs, though this got so out of hand one year that the computer had to be cordoned off from the hordes of youngsters eager to help themselves.

It was during those open days that Peter and Kay Thorne got an idea of how the public viewed computers. Peter reckons, "The public was mystified. People occasionally would come up and ask for answers to quiz questions or something like that. If they thought it was anything, they thought it was an electronic brain with infinite memory. They actually expected computers to be ahead of what they were—it was something like science fiction." Kay recalls the follow-up calls that resulted from the open days, "asking Frank to speak to all sorts of groups, from local women's clubs to a very senior company exec who came in for private training in the evenings for a few weeks so he would be able to understand the coming technology and hold his own at board meetings."

Some telephone calls were harder to deal with, says Kay Thorne, "usually starting 'I saw the computer there last open day and would you ask the computer. . . .' I lived in fear of these because it was hard to explain why not and a bit churlish to refuse to help. Of course, many simply needed an encyclopedia and I got to be quite a dab hand at referring people to the right section. One day Frank, who was a softie with things like that, agreed in a weak moment that we would try to find a crossword answer to which the clue was "a Philippine shrub in six letters," and the caller was sure it started with an A. The *Encyclopaedia Britannica* failed me and the university library was not hot on gardening books. I had to admit my failure— which they interpreted as the computer's failure." Such callers were simply ahead of their time; now the answer can be found in minutes on one of many dedicated crossword-solving websites.

Peter Thorne was one of those whose experience with CSIRAC led him into a central role in Australian computing, a professorship, and a long-term effort to preserve both the machine and its history. It was a stimulating time. "I remember having discussions with Pearcey about artificial intelligence, about how big a computer would have to be at current technology to have similar computational powers to the brain. We knew we were in on something exciting and new, but we couldn't imagine it would go as far as it has gone in terms of speed, small size, and huge amounts of disk storage. The disk on CSIRAC was about 3 kilobytes, it was 1,000 20-bit words, and the idea that one day you could buy 80 gigabytes of disk for less than £100 would have been just unbelievable."

With such expertise demonstrated from the late 1940s onwards, did Australia really miss an opportunity to establish its own computer industry? Harry Messel is characteristically certain: "We sure did, because the SILLIAC was built by Standard Telephones and Cables [STC] right here in Sydney, and it was one of the leading machines in the world. We could have gone on and built these things, but the government didn't believe in computing and it was only four or five years after that the government realized the potential, right? So it was a lost opportunity in Australia for computing; that's absolutely so. You know, it's a sad old story in Australia that they don't believe in the capabilities of their marvellous young people." But also looking back, Peter Thorne isn't so sure that Australia had any opportunity to build and export the big early valve-based computers. "My view is that CSIRAC kick-started Australia into the digital age around 1949–1950 and it started an interest in software. The idea of building hardware in large quantities in Australia when most heavy goods went by sea was probably a bit of a pipe dream. The future was probably more in building software than hardware. Quite a lot of people cut their teeth on CSIRAC or on SILLIAC and we've done pretty well in the software business ever since. It fits our education—our national culture almost—and Australians are pretty well represented around the world in the software business, so I don't think it was a loss. I think CSIRAC played its part." As, of course, did SILLY YAK.

CHAPTER 8

WATER ON THE BRAIN

Bill Phillips demonstrating his Hydraulic Economics Computer in 1950, with cigarette, as always, in hand. It helped propel him from a poor degree in sociology to a professorship in economics in just eight years (photo courtesy of the London School of Economics).

At Cambridge University there is a much-loved computer that looks like no other there has ever been, and it is still in use today. Like all the early computers in this book, it uses valves and tubes—but not electronic valves and tubes. These are mechanical valves and clear plastic tubes carrying brightly colored water round a network of pipes and tanks that graphically illustrates the flow of money in a modern economy. It is the Phillips Hydraulic Economics Computer as it was known in the UK, or the "Moniac" as it was later christened for export. Moniac was said to stand for Monetary National Income Automatic Computer, though the abbreviation probably came first, with its echoes of the ENIAC. The water is pumped by a war-surplus hydraulic motor from a Lancaster bomber, water levels rise and fall to represent savings levels and the like, interest rates and tax rates are represented by settings on the valves, and so on. But this is not just a model of the economy; it really is a computer. For example, it can calculate the effect of raising interest rates on savings, consumer spending, and the money supply; and it abruptly settled at least one long-standing dispute between Keynesian and Robertsonian economists.

It is so visually striking that more than half a century after it was first unveiled it reappeared in the great Venice Biennale arts festival of 2003 as the centerpiece of New Zealand's exhibit. This linked the Moniac with another curiosity, the Trekka, a 1960s Land Rover look-alike intended to give the New Zealanders their own car industry. The Trekka was powered by Skoda engines from Czechoslovakia, secretly exchanged for dairy products across the Iron Curtain. What the artist Michael Stevenson wanted to show was that "linked, the Trekka and the Moniac become a wishful mixed metaphor: the Trekka powering the national economy."

Whatever wishful thinking surrounded the Trekka, there were no illusions about the usefulness of the Moniac in its day. Now, as the water pump runs up to speed (the only use for electricity in this machine is to pump the water around the maze of pipes) and you watch the "savings" tank fill up, see the "tax rate" valve siphoning a proportion of income into government revenues, and watch the pens attached to various parts of the system charting the changing economy, you cannot help wondering what kind of mind could conceive of this and, what is more, actually build it?

Bill Phillips was born in New Zealand in 1914. As a young man he soon learned to fix things. There weren't enough people around to be calling on tradesmen whenever something needed to be repaired, so New Zealanders were pretty self-sufficient people. While still at primary school he salvaged an abandoned truck, fixed it, and drove it to school every day until the authorities got wind of this and put a stop to it. At the age of 16, while working on a "boring" hydroelectric project, he started weekly film shows for the workers and had to go to some lengths to avoid ever meeting the distributor, who had no idea he was dealing with a teenager. In his early twenties he took a route followed by many young New Zealanders, going west to Australia, where he bummed around for a couple of years, doing various odd jobs, including crocodile hunter and electrician in a gold mine. In 1937 he decided he ought to get some proper training and left for England. One day into his voyage to China on a Japanese boat, Japan invaded China, but he survived, arriving on an enemy boat and made his way to Russia and then, via the great Trans-Siberian Railway, to Britain in 1938.

He just had time to complete a correspondence course in electrical engineering before war broke out. He enlisted in the Royal Air Force and in the splendid irony of the services was sent straight back to the Far East. When Singapore fell to Japan, he was among those who got away on a troop ship, the *Empire State*. It was attacked from the air and his response was characteristic. According to a citation some years later:

> He obtained an unmounted machine gun, quickly improvised a successful mounting and operated the gun from the boat deck with outstanding courage for the whole period of the attack, which lasted for 3½ hours.

He survived and reached Java, but soon it too was overrun by the Japanese. According to a friend of his in later life, Professor Richard Lipsey, he still didn't give in easily. "Before they were captured a group of them disappeared and in an attempt to escape they found a bay with an old abandoned bus in it. They filled in the windows, put a sail and a keel on it, and they were going to sail away to Australia to escape. Perhaps fortunately the Japanese discovered them before they went off on this crazy enterprise and they were captured. I suspect it is apocryphal but it's an awfully good story and typical of Bill—he wouldn't give up; he would find some crazy solution and give it a try."

Phillips ended up in one of the notorious Japanese prisoner-of-war camps and for three and a half long years he, like many of his fellow prisoners, tried to make what he could of the grim conditions. There were many Chinese prisoners and he learned to speak their language, an interest that remained with him throughout his life. In spite of the sparse resources he found ways to apply his inventiveness. He designed a simple heater coil that could be used to boil up a single mug of water for a late-night cup of tea. It plugged into the crude lighting system, and so many were made by his grateful fellow POWs that it is said the camp lights used to dim as the bedtime cuppas were brewed. The guards never did work out what the problem with the lights was.

Like many Far East POWs, he almost never talked about his experiences, but other accounts surfaced over the years. The writer Laurens van der Post was a fellow prisoner and remembered him as the "gifted young New Zealand officer" who took their broken radio and repaired and miniaturized it, using parts filched from the commandant's office radiogram. It was designed to be hidden in a hollowed-out chair leg and on one nerve-racking occasion a guard conducting a search rocked back and forth on the flimsy chair. Phillips would have been immediately executed if the radio had been found, but he kept it hidden, until one day he heard a faint crackly voice saying that a terrible new weapon had been dropped on a Japanese city and the war would soon be over.

Repatriated to England, he resumed his studies, enrolling for sociology as he had become fascinated by the way the camp inmates had organized themselves. But the course was a disappointment to him and he did badly, barely passing the first year. However, all the sociology students had to do a supplementary course on economics and this, to his surprise, fascinated him, particularly the theories of John Maynard Keynes. According to Professor Nicholas Barr, a later contemporary of Phillips who is still at the LSE, "Bill Phillips found economics very difficult, but he was intrigued by an analogy in a textbook which compared monetary flows with hydraulic flows. This gave him a clue and he started translating economics into hydraulics."

One of his fellow students, Heather Sutton, remembers that it was like a revelation to him. "We were walking round Lincoln's Inn Fields one day and he explained to me that he had realized that economics was just simple hydrodynamics and he proceeded to wave his arms around and tell me about it. Not that I knew anything about hydrodynamics! But he really engaged my attention and I never forgot that meeting."

Phillips was using his engineering training to make the analogy between control theory and a modern economy, and Richard Lipsey says this gave him an advantage over staff whose background was pure economics. "He heard the staff fumbling around with Keynesian economics, in the sense that nobody really knew what the model was at that time. With his engineering background Phillips could see that they were playing with the kind of models he was very familiar with. Of course, others must have had the same experience, but Bill had this enormous insight to put the Keynesian concept together with the formal experience he had and out of this came the idea of the model."

Phillips had befriended an economics student a year above him, Walter Newlyn, and it was he who helped Phillips understand the economic theory he was struggling to grasp. Newlyn graduated from the LSE to take up a lectureship at Leeds University, while Phillips went into the final year of his degree. One day Newlyn was in London for a meeting at the LSE and the two men met up for lunch. Phillips produced a paper he had written for his coursework, describing how a hydraulic machine could be used to demonstrate and even compute such mechanisms in the economy. He sketched the machine and details of how it would work.

Some 55 years later, in the summer of 2001, an aging and infirm Walter Newlyn lit up with enthusiasm as he produced that original paper and recalled what an impact it had made on him. "I was carried away with the fact that here was a way of showing how the economy worked. The simile of using a water-flow diagram in economics had been familiar for a long time, but none of them had the features that Phillips's had. For the first time he could show not only the movement but the way in which the complicated interrelationship between two variables could be modeled. It was only one of the sectors—savings, investment, and interest—and this showed the thing that I went wild about. It showed an interrelation that would be dynamic and could therefore show the actual ongoing change resulting from any change in any of the variables, as well as the result of that change on any of the other variables through time. That was the fundamental achievement of that drawing, and as soon as I saw it I knew it was important. I asked him if we couldn't make something that actually did this and he said yes, it wasn't impossible." Until that point Phillips had treated the paper as a theoretical exercise, but now the two men decided they could actually build the machine, Phillips providing the engineering expertise, Newlyn the economic theory.

Before long Phillips found himself in Leeds, talking to Newlyn's head of department, Professor Arthur Brown. "Before we went to see the Prof I had converted the one sector that Phillips described in his paper into a complete model of the economy and it was this model that we showed Brown. He was very excited about it and gave him some money to carry him over a bit." It was £100, equivalent to over £2,200 today. Phillips's first approach to the head of his faculty at the LSE, Professor [later Lord] Lionel Robbins, was met with rather less enthusiasm. Robbins was skeptical but he did refer him to Professor James Meade and this time Phillips got lucky. Meade was a brilliant economist who would win a Nobel Prize for his work. He was also what one might now call a "gadget freak." He found the idea of a hydraulic computer irresistible and contributed both ideas and encouragement to the young man. He told Phillips that if it worked he would arrange for him to demonstrate it at a Robbins seminar—this was the high point of the week for the whole college, staff and graduate students in particular. The Robbins seminars were prestigious events that could make reputations or break them.

Newlyn recalled becoming rather worried when Phillips only got a third in sociology. He thought it was his fault as Phillips had spent most of the vacation before the exam working on how he would actually build his model instead of working for the exams. But with his degree out of the way, Phillips took over the garage in the house in Croydon where he was lodging and started work. Later he was to pay tribute to his long-suffering hosts, Phyllis and Bill Langley, who encouraged the two young men in their seemingly preposterous plan. Mr. Langley was a former Water Board engineer with a well-equipped workshop in his garage, so his home was an ideal base, and Newlyn remembered it as a great time. "Bill [Phillips] was an incessant smoker but he wasn't a workaholic and I admired him very much. He had a good sense of humor and was nice to work with. I was the boy as far as the mechanical work went, but I was the guide to some extent as far as the economics went, because his knowledge of macroeconomics was rather weak. He had only done a subsidiary course, so I had to fill in the gaps, but the ingenuity he used in making that machine really was incredible. He was a very inventive man."

By the end of the summer vacation the machine was complete. A large tank of colored water at the base represented the money supply. This was pumped up to the top, where it became income, and it then cascaded down through nine valves, each representing a different economic function. One siphoned off a proportion as taxes, another savings, and so on. So, for example, as the water in the tank marked "domestic expenditure" rose, a float rose with it, opening, via pulleys and cables, the valve marked "imports" and closing the valve marked "exports," thus showing how a spending spree at home can suck in imports from abroad and reduce the level of exports. The machine was very sophisticated, in effect constantly solving nine simultaneous equations and displaying the results in a form far more graphic than numbers on a page could ever be. Indeed in some ways it was better than solving equations, as it showed the transition from one steady state (say 4 percent interest) to another (say 5 percent interest) rather than just the final result. Phillips later added a system of charts and colored pens that automatically drew graphs of these relationships.

With the machine working, James Meade honored his promise and arranged for Phillips to demonstrate it at a Robbins seminar, which was one of the most crowded ever, due mainly to the novelty of the machine. The audience was skeptical to begin with as Phillips was a new graduate—and in sociology rather than economics. As Richard Lipsey was told when he joined the LSE shortly afterwards, "The economics seniors came along to have a bit of sport with the young upstart New Zealander with a poor degree in sociology and some machine that he claimed would model an entire national economy. Some of the junior staff told me he brought his landlady along in a great flowered hat! She'd been helping him as he didn't have much money to finance it, because she thought whatever he was doing must be brilliant. That was very typical of Bill. Particularly in those days when class mattered, most people would not have wanted to be seen in this august company with their landlady talking with the wrong accent. But not Bill."

That would only have encouraged those who, Lipsey was told, had come along to jeer at the machine: "Everyone was there to sharpen their knives and take this upstart apart. He set the machine up and started to talk. There were a few biting questions from junior staff and within five minutes they started going quiet. Bill had shown in just a few minutes that he understood how to model this kind of thing better than any of them, so they better shut up or show themselves up as not knowing as much as this third-class sociology student! So he gave this great performance for the best part of an hour to a quiet audience and they started asking him really deep questions and he answered them. Pretty soon he was lecturing them on how to relate what they were doing to what he was doing. I think he was pretty proud of that. He didn't have a nasty bone in his body and he never told me how he crushed those young guys, but several of the other staff members said it was a pretty good job in showing up the arrogant 'Young Turks.' "

Nicholas Barr, too, was told by James Meade and Lionel Robbins that in the process of demonstrating the machine Bill Phillips gave one of the best lectures on Keynesian economics that any of them had heard. As a result he was offered a job in the economics department, something virtually unheard of for a new graduate with a poor degree in a different subject.

The demonstration showed that the machine was a wonderful teaching tool and that Phillips was a natural teacher. Nicholas Barr says it was "the most extraordinary visual device, large, heavy, bulky, and also leaky, so there was good news and bad. When working, it was a marvelous teaching device, enormously illuminating in terms of the instant explanation of macroeconomics it conveys."

James Meade was delighted with his protégé, as Richard Lipsey saw during his time at the LSE. "Meade was a great Nobel Prize winner in international economics and he held graduate seminars and within the first two or three classes he unveiled the famous machine. James really thought it was wonderful in two ways: it really did capture a great deal of the economy and it was a great toy. You could see the child in James, like here's a train set we can play with and learn something too!"

Even Walter Newlyn was amazed at what he and Bill Phillips had built: "It really was marvellous. For people who were not mathematical, it was a miracle to see this kind of behavior before their very eyes." But it wasn't just a fine teaching aid, as Nicholas Barr found: "It was actually calibrated so that it was accurate to

±2 percent. It had an explicit underlying model, so if you asked what would happen if the Chancellor cut income tax by 10 percent, not only would it move from where the economy was to the new equilibrium, but the time path to the new equilibrium was accurate to ±2 percent. It had plotters on it that would record the time path [on graph paper] so it was quite extraordinarily accurate, but particularly the macroeconomic modeling, the prediction of what would happen if the Chancellor did x or the treasury did y."

This was something of a surprise to Walter Newlyn as he and Phillips had "actually set out to build the model as a learning aid, not as a computer, although some exercises in the economy were quite accurately run on the machine. No one would have thought of using it for computing anything except for saying this is accurate to a certain degree as a demonstration of how this relationship works out. But by the time Bill wrote his paper, which has a photograph of our model, the paper had run away from a teaching device to being a mathematical model and that is where Bill's expertise eventually developed." That paper, which described the machine they had built and how it accurately modeled the main relationships in a national economy, cemented his reputation as a bright new economist doing really original work.

The reputation of the machine (and Phillips) was further enhanced when it was used to resolve several major economic controversies of the time. Nicholas Barr says, "There are a lot of intellectual controversies which arise from looking at the same problem from two different viewpoints, and no one has put together the common denominator that makes it clear they are two sides of the same coin. The Phillips machine was brilliant at doing that, so it was realized that all sorts of things that had been major controversies in the profession shouldn't have been; they just depended on how you formulated the problem." The best known of these was a long-standing dispute between two great economists, John Maynard Keynes and Dennis Robertson, over the relationship between savings and interest rates in an economy. The Phillips machine showed both explanations worked and that "Keynes and Robertson need never have quarreled if they had had the Phillips machine before them."

As the machine's reputation spread, there was growing interest from other institutions in acquiring their own copy. Leeds took the first one, which they had largely funded, and the LSE commissioned a small engineering company to make a more sophisticated version, the Mark 2, which was taller, wider, heavier, better made, and included various extra facilities such as the graphical plotters, and about 14 were built. Several were sold for teaching purposes to other British universities, including Cambridge, Manchester, and Birmingham, and at least one went to Australia. A former LSE student, Abba Lerner, returned home to the US in the early 1950s full of enthusiasm for the machine, christened it the Moniac, and sold several more. One of these went to the Ford Motor Company, who used it to simulate the economy and try and predict car sales. The Central Bank of Guatemala bought one too, though this appears to have come to a sticky end. Graeme Dorrance, who was working for the bank in 1955 but under strict orders not to interfere with local staff, had to restrain himself as he saw it clumsily removed from the library and "was able to hear its separate parts descend the stairs noisily."

The Phillips machine's life as an economic forecasting model was limited because, as electronic computers developed, they were able to do the same job more accurately, more powerfully, and with less leakage of water. Indeed one of the enduring questions is why such a gifted electrical engineer chose hydraulics as the basis for his computer when so many other computing projects around the world were going electronic. Nicholas Barr researched Phillips's work for a paper in 1988, a project that allowed him access to many private documents. He found from his research that "at the time Phillips built it he *wanted* something visual. He was well aware of the emergence of electronic computers and indeed one of the things he did after building the machine was to learn about computers, working with the pioneers who developed electronic computers in Britain. He was thinking about the use of computers in economic forecasting in ways that were vastly more sophisticated than was possible with the simple visual machine. Anyone who has seen the Phillips machine will see that it still has the impact as a teaching device that it had 50 years ago."

This judgment is strongly endorsed by Dr. Brian Henry, who had a major role in refurbishing the Cambridge model, believed to be the only Phillips machine now in working order, still kept in a classroom in the economics department, and indeed still demonstrated on occasion to awestruck students. "I remember Phillips demonstrating the machine at the LSE when I was a student towards the end of the 1950s. It was the one he originally produced, a very modest affair. It was based on the same principles, of course, plastic tubes and plastic boxes to represent flows and stocks of money or income, valves to represent interest rates, the rate at which exports came in, and so on—the same principles but a much smaller machine on a pegboard. So it looked quite amateurish and there was this small man, about 5 feet 6 inches tall, bald-headed, always wore a suit, always smoking. And he demonstrated this to us students and I have to say for the first time—I was then in my third year as an undergraduate—I understood what was meant by the circular flow of income, and how the Keynesian multiplier worked. It was such a supreme visual telling of the mechanics of Keynesian economics that I think all the students began to understand for the first time what the basic ideas were all about."

At James Meade's instigation the LSE ordered two Mark 2

machines and this enabled some particularly spectacular demonstrations. For example, as Richard Lipsey recalls: "There were two machines representing two economies, the home economy and the foreign economy. And these two separate machines were linked together! Before that the exports would have just disappeared into a tank but now they led into another country, and the imports came from the other country, so these were money flows from us to them . . . and what happened in one affected the other. Then we had one guy as the Chancellor of the Exchequer who was juggling the tax rate and another guy as the Bank of England juggling the interest rate, which determined whether people spent a lot on investment or not. So you had the Chancellor doing something and the bank doing something else and you would see the inconsistencies! Since they all happened with a lag, which was one of his key points, it showed that running the economy is not easy. You do something today and you see the effect a year later, and maybe the Chancellor hasn't noticed what you're doing and he's doing something else. Then you throw the other two guys in from the other country and you begin to get an enormous feel of how complex and interrelated things are and, after all, compared to the real economy, this was still an awfully simple version."

Phillips rose rapidly from assistant lecturer in 1951 to professor in 1958. Nicholas Barr became one of his students later in the 1950s, when he was "very august but very friendly and approachable, surprisingly shy, and an absolutely brilliant teacher. It was easy for students to get bogged down in the technicalities. As an engineer, Bill did the mathematics almost as easily as breathing, but it wasn't so easy for the students. He'd explain complicated mathematics, then he'd say, 'but don't worry about all that, what it means is . . .' and always focus on what it actually meant."

Phillips's machines saw long service in the LSE, enlivening generations of students with their vivid depictions of the economy and their predilection for spouting highly colored leaks. Even if the machine didn't leak, a careless lecturer could set the economic factors in such a way that the water would overflow one of the tanks, a dramatic demonstration of an overheating economy. Eventually Phillips tired of the SOS calls from his colleagues and the machines fell out of use.

In the meantime he had forged ahead with his novel work in economics to become one of the leading economists of the 1950s. He is probably best known for his work on the relation between inflation and unemployment and in particular for his definition of the "Phillips curve," as it was dubbed. This showed the tendency for inflation to rise as unemployment fell and vice versa. It was appropriated by some politicians as justification for deliberate policies of creating unemployment to tackle inflation, a fact that greatly saddened him. He saw his work instead as describing a mechanism that was a challenge for politicians to engage with—to maintain full employment while restraining inflation in other ways.

The label "Phillips curve" was applied by others and there is a good case for saying that it should be known as the "Brown curve"— Brown being the Leeds University professor who had first put up money for the Phillips machine. One of Brown's most important works was *The Great Inflation 1939–1951*, which showed the inverse relation between inflation and unemployment, an empirical finding that pre-dated Phillips's more theoretical treatment.

For all the eminence that Bill Phillips achieved in his time at the LSE, he seems to have remained restless and dissatisfied. Heather Sutton recalls a man who didn't quite fit in to academia yet had some unusual personal qualities. "I first met him at an NUS [National Union of Students] conference and he told me about someone he thought I would like to meet. He brought her to my digs—in those days all one had was a small room and a gas-ring as a student—and this was Maud Geddes, another New Zealander, who became a very good friend. Bill's sensitivity kind of understood that we would get on. You don't often meet people with that rare gift of empathy." But she also remembers Phillips as someone who "people liked to be a bit scornful of, the kind of man they would like to put down, although he was unputdownable." It's something Sutton, who grew up in the West Indies, blames on the public-school culture of many of the staff and students, in contrast to the colonial background that she shared with Phillips and Geddes.

Richard Lipsey was close to Phillips for some years and when he moved to the University of Essex he tried to take Phillips with him. "Bill was very frustrated at the LSE. One of the problems with British universities is that they kill their professors with administration, and as soon as you get to be a professor because you're a famous researcher then they do everything they can to stop you doing research. LSE was an oligarchy of professors, run almost entirely by them. So for junior staff it was wonderful, there was almost nothing to do! But Bill was overcome by administration and had trouble saying no. So I tried to get him to come to Essex to a research chair where he would do no administration. But in the end his allegiance rested with the LSE and not long after that he decided to go back to Australia. And he never had as much of an impact as he should have, because he wouldn't write. In some ways the LSE administrative burden was an excuse: it was easier to put off writing for a bit and go to another committee than try and finish it. He didn't write a lot; he found it difficult. He liked to talk and play with ideas. I guess building the machine was easier than writing about it. He only did about a dozen papers, which for a man who's left his mark on the profession is a remarkably small output; today you'd expect 200 papers from somebody like that. He left several papers unpublished. He just didn't find writing easy."

In 1967 Phillips returned to Australia, partly so that his wife and children would be nearer their relatives. He took up an economics post at the Australian National University, but on the unique condition that he could spend half his time on Chinese studies. Intriguingly for a man so gifted in engineering and economics, it was Chinese that seems to have given him the most satisfaction. A couple of years later his wartime deprivation and the chain-smoking habit it had left him with caught up with him. He suffered a serious stroke, gave up his post, and returned to Auckland, New Zealand. Six years later, in March 1975, he died at age 60.

By the mid-1970s it appears all the Phillips machines had been retired to storerooms or basements. Nicholas Barr first got involved with them in 1972, when he was a junior lecturer at the LSE and started to revamp one of them. "I became fascinated by it and loved learning about it. In the later 1980s Tony Atkinson at the school wanted to restore it properly. He'd been exposed to the Cambridge machine and, realizing the machine at the LSE needed restoration, he got together funding to pay for it and asked if I'd like to write a paper about it. I thought it would be fascinating and what made it even more wonderful was that James Meade, who had given such strong backing to the young Phillips, had deposited his own papers under seal. When I approached Meade for access, he agreed, and so I drew on his personal papers and saw this wonderful correspondence with Phillips. What emerged was a fascinating picture of the history of economic thought and a remarkable period in the LSE's history.

"One of the many pleasurable aspects of the story is the tremendous personal relationship that grew up between James Meade and Bill Phillips. When they first met, Bill Phillips was a wild, wacky undergraduate in sociology with some crazy idea for a machine and James Meade was a very eminent professor at the LSE. James Meade took Bill Phillips seriously, supported him building the machine, once he'd seen the machine in action acted as his sponsor to ensure a better machine was built for the school, and was instrumental in getting Bill Phillips his first appointment at the school. So James Meade was very much his sponsor, but when you read the correspondence between the two, although very formal as befits the usage of the times, there was this wonderful warmth and friendship creeping in, in an understated English way. Here was a marvellous personal relationship, but also a very productive one academically. In fact you can say that if James Meade had laughed at Bill Phillips the undergraduate, then Bill Phillips the great economist might never have happened."

Only in the 1990s, as a by-product of the revival of interest in computer history and the growing realization that many of the early machines were in danger of being lost forever, was there a serious attempt to save the surviving Phillips machines. Brian Henry became involved in a comprehensive refurbishment of the Cambridge machine, for the economics department's 50th anniversary. This was one of the larger Mark 2 versions, 7 feet tall and 5 feet 5 inches wide, installed in the 1950s. It turned out that the original coloring agent in the water was corrosive and much time was spent looking for a replacement as well as refurbishing the damaged parts. Several colored dyes were tried, but each lost its color as it was pumped around the system. Eventually a compromise was found, not as vivid as the original but good enough to show up.

Two machines survived in the LSE, where in the late 1980s they were carefully restored to full working order, to the point where they could be coupled together to re-create one of James Meade's favorite demonstrations. On September 19, 1991, one of the machines was formally handed over to the New Zealand Institute of Economic Research. James Meade was there, along with Walter Newlyn, by then in his late seventies. The LSE's intention was to keep the other machine, but the sight of the machines in action prompted Doron Swade to see if he could acquire one for the

Science Museum. Negotiations took some time, as some of the LSE staff were reluctant to let their much-loved machine go, though at the same time they recognized they had no long-term ability to preserve it and they did want to see it go on public display. In the end the Science Museum won the transfer in return for an undertaking to keep it on public display, a rare privilege when only 5 to 10 percent of the museum's artifacts are exhibited at any one time.

A major concern was the corrosive colored water in the machine, and this was drained prior to a process Swade describes as similar to embalming. It had been decided that the machine would not be displayed in working order, though a video of it in action runs next to the machine. It was ready in 1995 and on March 22 it was installed in the Science Museum's computing gallery. It remains unique, a computer that is not merely outside the development of digital computing but outside the mainstream of analog computing too.

Walter Newlyn long outlived Phillips, making a career in development economics and spending much of his time in Africa, though always based at Leeds University, where he became a professor in 1967 and set up the African Studies Center. When interviewed for this book, he said he had "felt very old for 10 years" and he died not long afterwards, aged 87, having lived a very full life. He always referred to the machine as the "Phillips–Newlyn Economics Computer" and there is some justification for that—his part has sometimes been overlooked (though he showed no resentment at that). Certainly it was Newlyn who suggested that Phillips turn an interesting essay into a real machine, and Brian Henry gives him a lot more credit, saying, "I do see a lot of what is in the Phillips machine as the joint work of Walter Newlyn and Bill Phillips, with Walter providing a lot of the economic input."

But Phillips's reputation is well deserved, with Nicholas Barr summing up the thoughts of many: "Bill Phillips was one of the giants that bestrode the profession in the third quarter of the twentieth century, he really produced path-breaking research that moved economics in a major way. He was also in a personal way very warm, friendly, quiet, with a wry sense of humor, and a wonderful teacher. He's one of those rare people who, when his name is mentioned to anyone who knew him, a warm smile crosses that person's face."

CHAPTER 9

IT'S NOT ABOUT BEING FIRST: THE RISE AND RISE OF IBM

The man who created IBM and equipped it to dominate the computer age, Thomas J. Watson Sr., shown sitting at the console of the company's first truly electronic computer, the 701 or "Defense Calculator" (copyright IBM, courtesy of the IBM Corporate Archives).

There's a pretty good chance that if you stop people in the street, in any country in the world, and ask them who invented the computer, most of them will say IBM. Yet the name doesn't figure much in most of these early stories, at least not until the mid-1960s. One exception of course is the Harvard Mark 1, Howard Aiken's design built for him and paid for by IBM, but that electromechanical device was a technological dead end. However, when IBM did wake up to the importance of electronic computing, it moved quickly, making the most of its huge financial clout, great technical skills, employee loyalty, and fiercely competitive sales force. The roots of all that lie much earlier in the century.

The invention of the mechanical tabulator in time to rescue the US census at the end of the nineteenth century was a huge step in the history of automatic calculation. Demand for the new machine grew rapidly and its inventor, Herman Hollerith, soon found himself not only with a growing range of clients for his Tabulating Machine Company but also with competition.

While Hollerith was establishing the tabulator industry, Thomas J. Watson started out in other lines of business, initially with conspicuous lack of success. First it was selling hardware, then sewing machines, then shares in a dubious building and loan company. Disenchanted with selling, he opened a butcher's shop, but soon ran out of cash and had to sell out. Then he had his first stroke of luck. While transferring his cash register to the new owner of the butcher's shop, he got into a conversation with the local salesman for the National Cash Register Company. This was one John Range, who, after much badgering (early failure hadn't blunted Watson's persistence), took him on as his apprentice.

It was 1895, Watson was 21, and he failed to sell any cash registers at all during his first couple of weeks, but under Range's fierce tutelage he learned quickly. The "Cash," as the company was generally known, was molded in the image of its legendary and tyrannical boss, John H. Patterson, who had drawn up a very precise manual for his salesmen. Soon Watson was following the manual faithfully as well as using all the company's favorite unpublished methods: these included intimidation, undercutting, legal threats, spying on rivals' clients, sabotaging their machines, and so on. While NCR at one time had more than 50 competitors, most were progressively closed down by the dirty tricks or just vanished in the face of the Cash's overwhelming monopoly. A few survived but with only a small corner of the market between them.

Watson rapidly developed the absolute loyalty to NCR that he would later demand of his own workforce at IBM. As Patterson approached a total monopoly in the sale of new cash registers, he realized that most of the surviving competition was coming from his own used machines, sold secondhand. So in 1903 he recruited Watson to set up an undercover operation that would destroy the independent used cash register businesses across the US, and he gave him a million dollars to do it. One simple method was to open a store right next to one of these independent shops and beat its prices, buying for more and selling for less. Watson didn't have to make a profit and he might even sympathize with his rival's plight, blaming the ruthless price cutting of new machines by NCR. Eventually he would buy out the failing business, often quite generously, and move on to the next town.

Meanwhile Patterson was showering rivals—and even their

customers—with breach-of-patent claims; they were rarely pursued to judgment, being merely a means of tying up the companies in legal proceedings and scaring their clients away. Some rival sales agents were quietly given a second salary by NCR so they would stop competing too hard. And so on. This was capitalism red in tooth and claw and if it apparently breached the antitrust laws, well, that was a bit of a new idea anyway. A few decades earlier a trust—a group of businesses that combined to dominate the supply of a single product—was seen as a good thing. Only slowly did the drawbacks that went with such a monopoly become apparent, and the Sherman Antitrust Act of 1890 was an attempt to tackle the problem. At first it was hard to prove in court, but in 1911 the US government successfully took action against both American Tobacco and Standard Oil. The following year, buoyed by those victories and supported by an electorate suspicious of the growing power of big business, it prosecuted NCR. In February 1913 Patterson, Watson, and a third man, Joe Rogers, found themselves with large fines and worse, one-year jail sentences. Some of NCR's practices as described to the court made them sound more like mobsters than businessmen.

It could have finished them all, but within weeks an extraordinary stroke of fate elevated the disgraced executives to hero status in media and public opinion. The three men had been released on bail pending an appeal. Early one morning in March, Patterson saw that exceptional spring rainfall was causing the Miami River to rise rapidly as it flowed through Dayton, Ohio, the home of NCR. He'd long warned the city it was unprepared for flooding, but to little effect. He got his executives together and in a quarter of an hour turned them into a disaster relief team. They were to gather food, medicines, blankets, tents, and drinking water, and the carpenters were to drop everything and build simple rescue boats. It was barely seven in the morning. By half-past eight, waves of water were sweeping through the town and the flood rose with frightening speed into one of America's worst ever.

Within a couple of hours the town was almost cut off, and thousands of people were seeking refuge at NCR, fortunately built on a rise. There they found an emergency village, put together with amazing speed. For the 90,000 made homeless, Patterson's men were the main source of aid; Watson was out of town but he, too, dropped everything to help. By the following day he had chartered a train in New York and filled it with emergency supplies. NCR's "crooks" became overnight heroes and before long a petition was circulating calling for them to be pardoned. Ultimately things turned sour for Watson, whom Patterson began to see as too much of a threat—too much like him probably—to keep in the company. The relationship broke down, and Watson left NCR without another job to go to.

He took a lot of valuable lessons with him. From Patterson he took the model of a company built up into a dominant position in its market by one inspirational leader who demanded absolute loyalty from his staff and in most cases returned it, along with decent wages and benefits. From Charles Kettering he learned the key importance of research and development in a company (Kettering later left NCR to found Delco, inventing the electric starter motor and turning it into one of the greatest car parts companies). From the trial verdict he acquired a strong aversion to getting caught. That was a vital lesson and his biographer, Kevin Maney, concluded that "if Watson had never been brought to trial, he wouldn't have become the kind of leader who could build IBM."

Before long he was recruited to manage the Computing-Tabulating-Recording (CTR) company, set up by another buccaneering entrepreneur (and notorious gunrunner), Charlie Flint. Flint had been putting together trusts, 24 in all, since the 1870s, when they were seen as a good thing, so he had no sympathy with the antitrust movement and the Sherman Act. But CTR wasn't one of his best trusts. It was an amalgam of around a dozen companies, one of the most important being Hollerith's Tabulating Machine Company. Flint had more than one go at sorting it out without success and by 1914 he was desperate, so it was fortunate for him that Watson had lost his job and turned up in his office one day. Something told Flint that the disgraced executive was the man to turn his company around, a true entrepreneur, and he made him general manager. Watson showed how much he'd learned by that time, involving people, managing carefully around the rival heads of the companies that made up CTR, and inspiring a sense of purpose and family.

It helped greatly that a year later the original antitrust trial, verdict, and sentences were set aside. It helped even more that Woodrow Wilson was much less concerned about such matters than his predecessor, President William Taft, had been. The government offered NCR a "consent decree" that was little more than a promise of better behavior in the future. Watson took a big gamble in refusing to sign even that, still believing he had committed no crime and that a signature would be an admission of guilt. But Watson was no longer part of NCR and the government had no stomach to pursue him further, so his gamble paid off. It was a Saturday night. On the next working day the CTR directors met and made him president of the company. The next couple of years were a difficult time to do business. America wasn't in the Great War yet, but its European market had plummeted, arms sales aside (one of the factors that brought America into what was essentially a European war in 1917 was precisely the economic damage the war was doing to US interests).

In spite of the difficult market conditions, Watson kept the company going, and even growing. From the start he had set about creating a great sales force, inspiring them with an evangelical fervor that owed much to Patterson's example. And despite the antitrust saga, his methods included enough hard-edged practices to keep the authorities forever on the alert.

In 1924 Watson renamed CTR the International Business Machines Corporation, or IBM. He had already used the name for overseas subsidiaries and he liked it better than Calculating-Tabulating-Recording, which was both too specific and not grand enough. It didn't sound like the great corporation he wanted it to be, whereas International Business Machines did. Next he had to focus on one line of business. Part of the reason for changing the company name was to get rid of those products whose time was past, or at least would never make IBM into a great corporation. For example, they made time recorders that workers used to clock themselves in and out of factories. There was not much potential for growth there, and the same applied to everyday calculating machinery like scales.

Tabulating machines were another matter; Flint liked to claim that they could actually think, and Watson could see the growth potential in "data processing" as he called it (probably long before anyone else). He dreamed up machines for printing railway tickets, for automated banking and so on, and told his engineers to make them. Constant innovation was what he'd learned from Charles Kettering, along with the importance of patents. So what patents IBM didn't already own through their own inventions, they tried to buy. It all helped to build a stranglehold on the tabulating machine business, as did their patented punched-card design. No competitor could use the same layout, and no customer was allowed to use cards made by anyone else; the tabulating machines were leased not sold, so it was possible to make such rules stick. Rather surprisingly this didn't bring down another antitrust lawsuit on their heads.

If Watson was a hard competitor he became a benevolent employer, and this was something he'd learned from the other major company established in Endicott, where IBM had its main factory. Endicott-Johnson Shoes was run by George Johnson, who was a quite untypical boss for America in that era, and more akin to the Quaker philanthropists in nineteenth-century Britain. Johnson believed that happy employees made better shoes, so he cut working hours, raised wages, and gave them free meals and medical care, as well as a host of other benefits. The unions didn't get even a toehold in his company; Johnson was giving the workers more than they'd dare ask for anyway. It took Watson a long time to agree that his rival for the Endicott workforce was onto something, but when the penny dropped he embraced the same approach. It was another step on the road to what became known as the IBM culture.

He had also come up with the motto "Think" while at the Cash. It was all his own and he was immensely proud of it, bringing it with him to IBM and making it the title of a company magazine in the 1930s. Any subordinate without the "Think" motto on desk or office wall was sternly rebuked (even today the word lives on in products like the IBM ThinkPad laptop computer). But tangible company benefits grew too. There were three company country clubs, heavily subsidized and open to all staff. And somehow during those years the IBM "uniform" of white shirt and tie became established. A new recruit who persisted in wearing colored shirts would soon find himself with a "gift" of white ones from his supervisor. It was a white company in another way too. Few blacks got jobs, except at the lowest grades, and there weren't many Catholics either. This wasn't unusual in pre-war corporate America of course, and the situation at IBM was transformed from World War II onwards.

On the other hand Watson didn't believe in large-scale layoffs. In the weeks after the Wall Street crash, as companies across the country made plans for recession, Watson encouraged his executives to plan for growth. There would be no cutbacks and spending on research and development would go up, with the intention of inventing new products and creating growth that way. He was "betting the company" on riding out the Depression with piles of stock ready to sell to a resurgent America. It was a bet that could bankrupt IBM if it failed, but Watson was convinced the upturn would come in time and put them in a strong position. There was a less rational factor too, his son later recalling that Jim Rand, head of IBM's smaller rival Remington Rand, once asked Watson Sr. why he was still hiring salesmen at a time when mass layoffs were the norm. Watson replied that he was approaching 60, when a lot of men take to drink or chasing younger women, "but my weakness is hiring salesmen, and I'm going to keep doing that." Watson's IBM was above all about selling.

As President Herbert Hoover made way for Franklin Roosevelt, Watson became a fervent advocate of the New Deal. There's no indication there was anything cynical in this, so maybe it was justice that it probably saved IBM, after several years when production far exceeded sales. The Social Security Act passed under the New Deal on August 14, 1935, dumped a huge load of mandatory data-collection and record-keeping duties onto industry. Companies responded by leasing lots and lots of tabulating machinery, mostly from the only company that had great stockpiles of the stuff, a full staff, and lots of spares. The act of faith paid off dramatically for Watson and IBM. Before long he was the highest-paid executive in America, on $365,000 a year, dubbed the "Thousand-Dollar-a-Day Man" by the newspapers. Just over 20 years after being forced out of NCR with a criminal conviction hanging over him, he'd overtaken his mentor, John H. Patterson. More to the point, he'd kept his loyal workforce together, even expanded it, during the grimmest years of the Depression.

The 1930s were important in other ways for IBM's progress towards eventual domination of the computer world. Watson got his first glimmer of the future with a call one day from Benjamin

Wood, an educational psychologist who needed help evaluating millions of results from tests he had devised to methodically assess student achievement (distant forerunners of today's SATs in schools). Wood had approached ten chief executives and Watson was the only one who would see him. Watson gave him an hour, but on the day it was five and a half hours before he let Wood out of his office. In that time Wood had opened Watson's eyes to the fact that almost every human activity was measurable, and that meant a potential market in almost every field of human activity for IBM machines. Within a couple of days Wood had the equipment he needed and before long he was opening Watson's eyes still further, saying that the machines would one day operate at the speed of light—electrically. But that was some way off.

First came the establishment of the Thomas J. Watson Astronomical Computing Bureau at Columbia University, thanks to another persuasive individual who managed to get Watson's attention. This was Wallace Eckert (no relation to the ENIAC's Presper Eckert). It wasn't computing as it came to be known in the electronic era, but the mechanical calculators donated by IBM were put to good use by Eckert. Watson was very proud of his astronomical bureau and his links with the university.

He wasn't so perspicacious in other ways. In 1937 he visited Germany, where Adolf Hitler awarded him the Order of Merit of the German Eagle with Star (the country's second-highest decoration). Hitler granted Watson a private audience and assured him that Germany didn't want war. Later Watson sent the Führer a warm thank-you note for his hospitality and he continued his European "world peace through world trade" tour by paying tribute to the dictator Benito Mussolini at a meeting of salesmen in Italy. On his return home he spread the message through "Peace" signs placed throughout IBM offices and factories next to the "Think" mottos. This was not Watson's finest hour. He wasn't the only one fooled by Herr Hitler at that time, and there's no evidence he actually approved of Fascism, but he failed to use his influence at the top of the German government to condemn or even restrain Nazi oppression. It was 1940 before he returned his medal, an act that sent Hitler into one of his rages, banning him forever from German soil.

Those public episodes are far from the whole story of Watson's involvement with Nazi Germany. IBM owned 90 percent of the German company Dehomag, whose Hollerith census tabulator was used by Hitler to create a new census after his accession to power in 1933. Tracking race and religion was a major part of this exercise and the machines helped the Nazis locate Jews, seize their businesses and property, and eventually herd them into ghettos or drive them out of the country. Dehomag machines were even used to find those individuals who had a single Jewish grandparent or had converted to Christianity, or otherwise weren't readily identifiable as Jews. Later, banks of tabulators would be used to keep track of prisoners in concentration camps, from admission through forced labor to extermination. Even after Watson returned his medal to Hitler, the Nazi concentration camp and forced labor system relied so heavily on tabulating machines that there was no question of the IBM-owned Dehomag losing its contracts, and revenues continued to mount.

The case against Watson has been meticulously researched by Edwin Black, who concludes that "IBM's business was never about Nazism. It was never about anti-Semitism. It was always about the money. Before even one Jew was encased in a hard-coded Hollerith identity [a punched card], it was only the money that mattered." Maybe it was more precisely about "selling" than money. It's hard to imagine Thomas J. Watson Sr. ever turning down an opportunity to sell, anywhere in the world. His eldest son, Thomas Watson Jr., claimed, "Dad's optimism blinded him to what was going on in

Germany" before the war, and it should be noted that he was just as positive towards trade with Russia. In the early 1930s he had risked serious unpopularity among business people by backing President Roosevelt, who was being criticized as "soft on Bolsheviks," and Watson Jr. reveals that IBM "did substantial business with the Soviets, who relied on IBM machines to manage vast quantities of statistics for their Five Year Plans." He also says that when he wrote to his father during the war about the appalling conditions he saw for himself in Moscow he got a reply that included the following advice: "You must keep in mind that every country is in a position to figure out what is best for its own people. It is not our duty to either criticize or advise them in these matters."

The importance of Hollerith tabulators to Hitler's Holocaust is undeniable and is well illustrated by two of many telling statistics in Edwin Black's book: in the Netherlands, where a willing collaborator, Jacobus Lentz, tabulated the census with great efficiency, around 75 percent of the Jewish population perished; in France, where the census was repeatedly subverted by a brave double agent, René Carmille, the Jewish death rate was 25 percent.

When Black's book was first published in 2001, IBM responded with a statement that included: "IBM and its employees around the world find the atrocities committed by the Nazi regime abhorrent, and categorically condemn any actions which aided their unspeakable acts. It has been known for decades that the Nazis used Hollerith equipment and that IBM's German subsidiary during the 1930s—Deutsche Hollerith Maschinen GmbH (Dehomag)—supplied Hollerith equipment. As with hundreds of foreign-owned companies that did business in Germany at that time, Dehomag came under the control of Nazi authorities prior to and during World War II. These well-known facts appear to be the primary underpinning for these recent allegations." When the paperback edition appeared the following year, IBM reiterated that position and pledged to "continue to cooperate and support legitimate research."

If IBM's wartime involvement with its German subsidiary remains controversial, there is little question that it did very well out of the war at home, thanks in large part to another huge gamble by Watson. Once the fiasco of the Hitler medal was behind him, he threw IBM into the war effort, risking the company again in an ambitious plan to use military contracts to continue growing. During those years he doubled the size and production capacity of the company, with the intention of maintaining that level of activity after the war was over. Laying off his loyal war workers in peacetime was unthinkable to him, and there were thousands of staff returning from armed service to find jobs for as well (they'd been kept on 25 percent salary throughout their time away). There was also a less tangible consequence of the war, the equivalent to people like Maurice Wilkes "learning how to get things done." Watson Jr. says that his father "somehow sensed correctly that, because of the war, the pace of technological change in American life had permanently changed."

Most commentators expected a post-war downturn in the economy and many companies assumed that any boom in war work would be a temporary blip. The prospects for IBM looked even worse as it still leased tabulators rather than selling them, so a flood of used machines were due back into the factory from the military and war industry alike as soon as peace was declared. This would depress the demand for new machines, just when IBM needed to keep production high. But Watson's fertile mind had an answer to that: he told his engineers to find ways of "detuning" the returned machines to run at just one-tenth of their design speed. These would be leased cheaply to smaller businesses that couldn't previously afford IBM, thus opening up a new market and simultaneously taking the used machines out of the existing market.

If there was anything unethical about this practice it doesn't seem to have occurred to him, although his biographer, Kevin Maney, calls it "a practical decision of murky integrity."

IBM had taken another important step during the war, although in this case they backed the wrong horse. Wallace Eckert's use of mechanical calculators in the Columbia Astronomical Bureau had attracted the attention of the Harvard professor Howard Aiken. He had his own ideas for a high-speed mechanical calculator and wasn't impressed with the relatively slow IBM gear at the computing bureau. But Aiken needed a patron and there didn't seem to be anyone else around, so he persuaded Watson to put up the money to build the Automatic Sequence-Controlled Calculator (ASCC). Almost as important, Aiken got access to the IBM engineers, including the remarkable James Bryce, whose hundreds of inventions included many components that were to become crucial parts of the ASCC.

IBM even built Aiken's machine in its own laboratory. It took four years and the original price tag multiplied several times over. Quite what that price tag was is open to question. Many writers say Watson put a million dollars up front and five times that by the time it was finished. Kevin Maney has an initial $15,000, rising to half a million. Whatever the true figure, it was finished in 1943, given months of testing, and sent to Harvard the following year. Aiken and Watson, two giant egos who by now could barely stand each other, had a huge row over its appearance. Aiken wanted all the workings open to the world, so his contemporaries could admire his ingenuity. Watson wanted a curvy glass and stainless-steel cabinet for his half-a-million-dollar baby, and he won. Not that "baby" is quite the right word for a 2-ton machine over 50 feet long and 8 feet high with three-quarters of a million components.

The night before the Harvard Mark 1 was to be unveiled to the public, Watson found out that Aiken was planning a ceremony in praise of himself and his university, with almost no reference to Watson, IBM, or the company's enormous contribution. Watson was all for reclaiming the machine and taking it straight back to IBM, and only a more emollient man than Aiken, the Harvard president, James Conant, managed to save the day with an impromptu speech praising IBM and its president. But the breakdown in the relationship between these two lumbering egos was permanent. It did at least provoke Watson into ordering his engineers to come up with a better computer than the Mark 1, a petulant act that in effect drove the company, hesitantly, towards electronics. Of course, Aiken quarreled with many of his contemporaries, as Maurice Wilkes attests. Though a university professor, Aiken at times seemed to despise his academic colleagues, reveling in provoking violent arguments with them. Still Howard Aiken was undoubtedly a great pioneer of the advanced mechanical calculator and at the very least he deserves some credit for prompting IBM to take its first faltering step into the new world of automatic computing.

IBM was strangely slow to join in the great drive towards stored-program electronic computers immediately after the war. Much of the reason for that is simply that Watson was getting old and no longer embracing innovation as eagerly as he had done for most of his working life. He simply couldn't comprehend that electronics both threatened his existing punched-card empire and at the same time offered a great leap forward into the next generation of data processing. There doesn't, however, seem to be much truth in the endlessly repeated claim that he said, in 1943, "the world will only ever need about five computers." If anything like that was ever said it may have been specifically about the predicted demand for Harvard Mark 1-type machines, and more likely said by Aiken than Watson.

Fortunately for IBM, Watson Sr. had long been desperate to pass the company onto his son, Thomas Watson Jr. This was shameless nepotism. In fact he believed in nepotism as good for business and he had even founded the "IBM Father-and-Son Club" in the 1920s, with the specific aim of encouraging family dynasties among employees. However, Watson Jr. was troubled and troublesome, perhaps because of the high expectations resting on him. Despite trying a succession of colleges, his academic achievements were meager and he preferred fast cars and marijuana. A spell at IBM before the war led to rapid promotion, but it was all based on favoritism and he knew he had no real respect from those around him. He left IBM, joined the US Air Force, and shortly afterwards got married. It is not clear which step matured him the most, but certainly his years as aide to the commander of the First Air Force allowed him for the first time to discover his real talent for organization and to demonstrate it to others. After demobilization, and after considering a range of options that did not involve IBM, he decided to rejoin his father's company.

He was made assistant to one of Watson Sr.'s most trusted executives, Charley Kirk, and one of their early duties was a visit to the ENIAC, a trip that he later admitted "could well have changed the course of computer-industry history, if either of us had understood what was in front of our noses." They were impressed by the great black 18,000-tube machine calculating the trajectory of a shell faster than it took to fly to its target, yet they failed to see its potential as business equipment. Years later Watson Jr. said he couldn't imagine why he didn't think "good God, that's the future of the IBM company."

Some weeks later, back in the IBM lab, he saw a demonstration of a simple electronic calculator made at home by an enthusiastic young engineer, Halsey Dickinson. It opened Watson Jr.'s eyes to the potential of electronics and later the same year the IBM 603 Electronic Multiplier was unveiled at a trade show. The Old Man wasn't sure anyone would want it, but customers queued up and the demand led rapidly to a more versatile 604. These were not computers, but at least they were electronic.

The IBM engineering lab was also working on the Aiken-beating "super-calculator," or Selective Sequence Electronic Calculator (SSEC) to give it its proper title, but the word "electronic" flattered to deceive; it thumbed its nose at Aiken's electromechanical Mark 1, but this was still a hybrid of electronics and punched-card technology. This one really did cost almost a million dollars and Watson was delighted with the SSEC, which he had finished in glass and shiny metal and placed on display, fully operational, in the ground-floor window of IBM's Madison Avenue headquarters in New York. But it was just another step down the same cul-de-sac inhabited by the Mark 1.

What's more, the company had a nasty shock in 1947 when the US Census Bureau announced it was buying two UNIVAC computers to replace its IBM tabulators, now regarded as too slow for the burgeoning census. It wasn't just losing the contract that was shocking, but the fact that it was the Census Bureau, the very organization that had prompted Hollerith to invent his tabulator for the 1890 census and in effect given birth to the IBM line. A line that had given IBM 80 to 90 percent of the data-processing machine market for over 30 years. The Census Bureau was buying the same machine whose creators they had dismissed the previous year and, worse still, more than a dozen other companies were building computers with government support of one kind or another.

In 1949 IBM passed up its second chance of acquiring the UNIVAC. When the Eckert–Mauchly Computer Corporation lost its chief financial backer, Henry Straus, in that fatal plane crash, Eckert and Mauchly went to IBM with a proposal to turn the UNIVAC into a joint venture. The Watson laboratory staff were enthusiastic. It seemed a perfect match: Eckert–Mauchly had the expertise and the design and had done most of the development, while IBM had the finances, the engineering skill to "productionize" the computer, and the sales force to sell it. It could have been called the IBM UNIVAC. But Watson said no, with the gnomic statement "no reasonable interaction possible between Eckert–Mauchly and IBM." Thomas Watson Jr. later reckoned this was simply because the lawyers had advised that buying the company would breach the antitrust law that still worried and angered his father.

IBM could have foundered in the 1950s, holding onto its monopoly of a technology that was becoming obsolete while failing to get to grips with the electronic computers that were threatening to replace its mechanical calculators. The scale of the challenge in 1950 was enormous, but by this time Watson Sr. was already more than 10 years past normal retirement date and Watson Jr. was being groomed to take over (he had been a vice-president for three years by now). Not only did IBM need Watson Junior's younger mind to embrace the new computing, he needed electronics to differentiate himself from his father.

Not that his father was against going electronic; in fact Tom Jr. said, "Electronics was the only major issue on which we *didn't* fight," but it needed the younger man's more positive attitude to the new technology. He had already shown his hand by taking control of the engineering lab. This had long been one of his father's most treasured parts of IBM, but the younger Watson had seen it clearly for what it was: a pre-electronic workshop filled with mechanical engineers (he dismissed them as "spanner-monkeys'). He put in a new engineering director with orders to recruit hundreds of electronics engineers (if Eckert and Mauchly knew, they must have been green with envy at such resources). As he put it, "From around 1950 my goal—one of the things we never saw eye to eye over—was to push into computers as fast as possible. The risk made Dad balk, even though he sensed the enormous potential of electronics as early as I did."

The opportunity to put the new capability to good use came with the approach of the Korean War. The military wanted computers for several distinct tasks and the IBM engineers proposed a single general-purpose programmable design. Tom Jr. agreed, calling it the "Defense Calculator" and Watson Sr. (still in overall charge as president and chairman) approved it.

Design work on the IBM 701, as it became known, started in January 1951. Costs spiraled upwards. This was a $10 million program, already 10 times the cost of the SSEC. Rental would cost their customers an eye-watering $8,000 a month, yet within a couple of months of starting (long before anything resembling a prototype had been built) they had 11 firm orders and 10 more good prospects. Just 18 months later they began building the first production model. Watson Sr. may have been slow off the mark, but he'd built his company well. The pride of his fiercely loyal engineers had been dented by the UNIVAC's growing prominence. The old days of planned schedules and detailed costings were left behind, and the project was tackled more like a wartime emergency. If a part wasn't in stores, they made do with what they could find. If progress was slow they worked late.

On January 15, 1952, Watson Sr. finally made his eldest son president of the company, though he remained as chairman and chief executive. It was a happy moment for both of them, but less than a week later the Department of Justice announced a new antitrust suit against IBM. The surprising thing perhaps is that it took so long. If the company lost it would be broken up and the 78-year-old Watson would go down in history as a monopolist, the label he'd struggled to lose 40 years earlier, so he wasn't going to settle with the government at any price. The young Watson saw the matter quite differently, wanting to settle not just to get the issue out of the way but also to make the company face up to the fact that the punched-card business was obsolete and no longer worth monopolizing anyway. They needed to move forwards whole-heartedly into the computer age.

Some time during that year the Watsons realized that the 701 was going to be much more expensive than they'd thought. Rental would be not $8,000 a month but $12,000 to $18,000. To Watson Jr.'s amazement, not one customer canceled on being told the news and he realized that "customers wanted computers so badly that we could double the price and still not drive people away."

Election night 1952 brought the biggest shock of all. The IBM 701 was just a couple of months from launch when Watson Sr. sat down in front of the television, expecting an enjoyable night seeing his old friend General Dwight D. Eisenhower narrowly beat Adlai Stevenson. Instead he watched with growing rage and disbelief as the UNIVAC triumphantly predicted the ensuing landslide and humbled the human pundits in the process. For some years to follow, the media and public were much more likely to talk of "UNIVACs" than "computers." In the ultimate humiliation, potential customers would refer to the IBM 701 as "IBM's UNIVAC," doubly aggravating when it could indeed have been the IBM UNIVAC if they had been more receptive to Eckert and Mauchly's work in earlier years.

It mattered little in the long term though as IBM quickly overtook UNIVAC. On New Year's Eve 1952 the first 701 was delivered to IBM's own computing bureau, and the first customer 701 went to the Los Alamos atomic weapons lab in May 1953. By early 1954 Remington Rand had 20 UNIVACs installed against 15 IBMs, but already IBM had four times as many orders. This is where the huge company built in Watson Sr.'s image had the advantage. Its resources were gigantic, its self-belief enormous, its engineering excellent, and above all it could sell. Moreover, it sold data-processing equipment and nothing else, while Remington Rand sold typewriters, razors, television, punched-card equipment, and more besides. And when it did sell UNIVACs, Rand couldn't cope with the demand anyway. IBM could supply their machine, which was also better styled and more practical—a collection of sleek metal cabinets the size of domestic fridges that were easy to deliver, connect, and get working. A UNIVAC was a behemoth whose first customer had preferred to leave it in the factory rather than risk rebuilding it at the Census Bureau.

It's ironic that the little company Remington Rand produced a monster of a computer, while the giant IBM designed a more practical collection of modestly sized cabinets. It's also a reminder that big organizations don't have to move slowly and a nimble company with massive resources is an almighty competitor. It's right, though, to recognize IBM's thoroughgoing professionalism: much of the early computer industry was characterized by amateurishness in hardware, finance, marketing, and business skills; IBM understood only too well that technology was about "turning inventions into products."

For four years the antitrust suit hung over the company and in particular the aging Watson, still refusing to "admit guilt" by agreeing to a settlement. Finally after a blazing row with the younger man, he left Tom Watson Jr. to agree a consent decree. It meant IBM would have to sell machines as well as lease them and relax their stranglehold on the punched-card market. It wasn't a bad outcome: these weren't the gangster-like practices of the earlier NCR trial, and besides it released the might of IBM onto the computer market.

For the rest of the 1950s, IBM had to cope with a range of competitors in the computing field, although there was great satisfaction when the next presidential election night in 1956 was covered live on TV with IBM computers, not UNIVACs. Still, it must have seemed like the old days of market dominance were slipping into history along with the old punched-card machinery.

By the beginning of 1961 IBM had its name on over 4,000 of the 6,000 computers installed in the US, and it was a similar story overseas. Peter Titman, a former LEO engineer who became a salesman for IBM United Kingdom, says that by the end of that year "we weren't the largest company in the UK, but in almost every other country we were. We were selling the IBM 1401 in incredible numbers. You could go to a company that had a punched-card installation, say two accounting machines and a calculator, and sell them a 1401 for the same price. They could do very much more with it, and they could do it simply."

Strong though IBM's position undoubtedly was, it still didn't match its past dominance of the tabulator market. However, before the end of 1961, IBM came up with a masterstroke, a plan for a range of computers that would form a "family." Smaller businesses would be able to afford the basic model, while a series of progressively more powerful models were built onto that base. The idea was that a company could buy the version most suited to their needs, secure in the knowledge that it would be easy to upgrade to the next model; all software that ran on the smaller one would transfer directly to the larger. Up until then buying or leasing a new computer almost always meant an enormous workload of staff training, software reprogramming, and so on. It was another plan that made use of IBM's size and resources; no other company could engineer such a range of compatible computers and sell and support them all. It would be called the 360, the number of degrees in a circle, symbolizing the claim that this computer would "encompass every need of every user in the business and the scientific worlds."

Visionary though it was, the IBM 360 was another huge gamble. Just as his father before him had twice "bet the company," once during the Depression and again during World War II, Watson Jr. now staked IBM's future on his $5 billion project. This was the first "third-generation computer" with integrated circuits replacing transistors, just as transistors had previously supplanted vacuum tubes. If it succeeded it would make IBM's own existing computers obsolete, as well as all the competitors' products. That was bad enough, but if it failed IBM would collapse, and it nearly did, thanks to the other Watson Jr. in the company.

Arthur Watson, always known as "Dick," was the youngest of the four children (there were two daughters as well) and often an unhappy one, desperate for the love and attention lavished on elder brother Tom. Dick had been brought into IBM at a high level and put in charge of World Trade, which comprised all the overseas companies. Relations between the two brothers were often strained and Dick's performance didn't always help, though overall IBM World Trade was thriving. So in 1963 Tom made him head of Engineering and Manufacturing, with the IBM 360 his major responsibility. The project went badly: 360s due for delivery in spring 1965 didn't materialize, software didn't work as intended, and compatibility wasn't as complete as promised. Gleeful rivals staked out the moral (and practical) high ground by advertising that they only promised what they could deliver. Tom demoted Dick and put Vin Learson in charge—simplifying the product line had been Learson's idea in the first place and Tom regarded him as "the father of the new line of machines." It was a move that broke Dick but saved the project and probably the company. Soon the 360 series was a runaway success, and rivals scrambled to produce their own "families." In 1966 Learson became the company president, and by the end of the year IBM had between 7,000 and 8,000 System 360s installed, a staggering quantity in less than two years of production. By comparison the British LEO sold barely 100 systems in its whole lifetime.

By the late 1960s IBM was on its way to dominating the computer market as it had so long dominated the punched-card tabulator market, and would do so for another 20 years. Writers like William Rodgers, who wrote a classic critical history of IBM in 1969, saw even then that computers would "become more readily adaptable for use by people with only marginal training in technical aspects of their functions. Lawyers, architects, run-of-the-mill engineers, middle management men, and, in due course, ordinary households will be encompassed by the computer market." It's both impressive foresight on Rodgers's part and a reminder of how far we have come that such use is now commonplace.

In Britain LEO was no more, with ICT (International Computers and Tabulators) and English Electric propped up by government subsidy to try and hold back the IBM tide; one of the most talked-about books of the day was *The American Take-over of Britain* by James McMillan, in which IBM was the lead villain. In the Soviet Union, a political decision had been taken to abandon most indigenous design and copy the IBM 360 architecture. Australia had long since thrown away its early opportunities of establishing a native industry. Across continental Europe, IBM had at least half of each country's computer business, up to 80 percent in some. No wonder someone coined the name "Big Blue" for IBM, whose salesmen and women could always quietly reassure a wavering customer that "no one ever got sacked for choosing IBM."

EPILOGUE

One of the recurring themes in these stories of the early computers has been the effect of World War II. Some projects were either a product of the war effort or at least advanced by the military experience of the people involved, while others were undoubtedly hindered by it. John Atanasoff and Cliff Berry's ABC is a good example of the latter. Atanasoff conceived the design before the war and it owed nothing to America's growing war effort, which eventually put an end to the project. Without the war they would surely have continued development of the ABC into a fully working prototype and ensured it was patented. It might not have developed at any great pace and would probably have been of most interest to other academics. It might even have fizzled out, particularly given Atanasoff's later lackluster performance as head of the Naval Ordnance Laboratory's computer project—maybe he was a true successor of Babbage in that respect. However, it seems more likely that Atanasoff and Berry together would have made something of the ABC, perhaps developing a niche market for small scientific computers, but it's hard to see it producing the kind of explosive growth that the ENIAC gave rise to.

It is very unlikely the ENIAC would have been built without the army's support. There was little prospect of building it as a weather-forecasting computer (John Mauchly's original intention) and no indication that even a company the size of IBM would have been willing to risk so much money on a machine with such uncertain prospects. The ENIAC needed a pressing military demand and the sort of funds that only a nation at war can find. Even after the war, military contracts kept it going and again it wasn't just a matter of hardware. Many of the engineers who formed the core of the post-war computer economy were trained directly on the ENIAC, learned related skills in wartime service, or acquired a technical education through the GI Bill after demobilization. Without the ENIAC there would have been no Moore School computing course in the summer of 1946, or all the projects that flowed from it at home and abroad, and it's difficult to see how the EDVAC or UNIVAC would have come about. If anything like the UNIVAC had emerged it would surely have been many years later, perhaps to star in the 1960 election instead of the 1952 one.

On the other side of the world the Soviet MESM was certainly driven by national security needs. It too benefited from an increased military budget sustained by the opening years of the Cold War, by the development of war technologies, and by the experience of war-hardened engineers. If the memories of Sergei Lebedev's widow are correct, he might well have started building an electronic computer earlier in the 1940s had the country not been at war, but it is doubtful he would have had the resources he was able to gather at Feofania, particularly as the public unveiling of the ENIAC at the start of the Cold War had a lot to do with generating those resources. It's much more likely that in the absence of war (and hence of the ENIAC) Lebedev would have worked on a much smaller computer over a longer timescale, much as John Atanasoff and Cliff Berry were doing a few years earlier.

Possibly something like the EDSAC would still have been built at Cambridge, but this time based on the ABC, if that had indeed been followed through. Had Atanasoff and Berry shared their knowledge, then Maurice Wilkes might still have built his own version, which was a natural step for the Cambridge Mathematical Laboratory. Or, if the ABC had become a production model aimed at other universities, surely Wilkes would have ordered one. His primary objective was not to build an electronic computer but to provide an up-to-date computing service for the university.

Even the LEO, a business computer with no military uses, was to a considerable extent a product of World War II. LEO met a need that was suddenly brought into sharp focus by the reduced availability of clerks immediately after the war and the economic demands on the teashops, with their tiny profit margins. LEO in turn was based, via the EDSAC, on the ENIAC, and it too relied on technologies that were greatly accelerated by the war, such as the mercury delay-lines that were at the heart of the design.

There are many other examples, most of them illustrating the impetus the war gave to the development of electronic computing. The Rand 409 is rather different and gives another clue as to how computing might have developed in the absence of war. Its creator, Loring Crosman, dreamed up his idea for a business computer around 1943 and it seems to have had no relation to the ABC or any other electronic computer, or the war effort. Rather his idea was for an "electronic punched-card computer" (as the Rand 409 sales brochure described it)—taking the mainstay of office calculating equipment, the punched-card tabulator, and building an electronic computer around it. It had no memory to store programs and it couldn't do very complex computation, but it was quite capable of basic payroll work and tax calculations. It wasn't particularly expensive, either to develop or to purchase, so it didn't need military budgets to support it. With around 1,500 sales by 1960 this is the most likely model of 1950s computing in the absence of World War II.

The Rand 409 would not have been the only design around in that event. As well as a completed ABC there would probably still have been Konrad Zuse's Z-4 and Howard Aiken's Harvard Mark 1, both electromechanical computers. None of these machines, though, had stored-program memories, the major innovation that emerged in the immediate post-war years and was the basis of the modern electronic computer. It's unlikely that either the technologies or the pressure to develop stored-program memories would have been there in the late 1940s. It would have been a longer-term development, perhaps spreading across the 1950s. If the Rand 409 is a good indicator of possible computer development in the absence of World War II, then it's significant that the Rowayton team's next model, the UNIVAC 1004, a brand-new 1960s design for the transistor age, still had no stored-program memory in its first incarnation.

By the 1960s, though, the influence of World War II on technology was declining, and it's certainly arguable that even if the war had never happened the 1970s would have experienced much the same rapid advance in computing technologies, perhaps just delayed by a few years.

The LEO name disappeared after the series of mergers in the 1960s that culminated in the formation of English Electric Computers, whose main competitors had already come together as the International Computers and Tabulators company. ICT, which included the computer divisions of GEC, EMI, and Ferranti, was pushed by the British government to merge with English Electric and create International Computers Ltd. ICL survived a couple of decades, never quite fulfilling its intended role as an international competitor to IBM, and was eventually taken over by Fujitsu, which still has an important place in the British computing industry.

There is a physical legacy in the form of those early computers or parts that have survived and are now carefully preserved, occasionally even in working order. Nothing much remains of the first LEO or the Cambridge EDSAC it was based on, but several complete units from the American ENIAC are still on display at the University of Pennsylvania.

Not a single example of the 1,500 Rand 409s (most badged as UNIVAC 60/120s), manufactured during the 1950s and delivered all over America, is known to have survived. Most of the few parts that remain owe their continuing existence to Remington Rand engineer Cliff Beierle, who used to take bits of scrap and turn them into presentation pieces for colleagues on their retirement. He saved a lot of components and chassis destined for the crusher, and a few of them have now found their way to the Rowayton Historical Society. Only in recent years, thanks largely to Erik Rambusch and Bill Wenning, have the engineers realized how significant their work was and set down the history whose records were almost entirely in their own memories. As Michael Norelli puts it, "You have to look back to realize what you did."

The original Atanasoff Berry Computer was broken up shortly after World War II, but the reinstatement of John Atanasoff and Cliff Berry to a proper place in computing history led to a campaign to rebuild their computer. The ABC has been painstakingly reconstructed at Iowa State University (formerly Iowa State College) in Ames, where it is occasionally shown to the public.

The most unusual machine in these pages, the Phillips Hydraulic Economics Computer, or Moniac, has survived rather better than most despite, or maybe because of, its mechanical nature—unlike those of its electronic contemporaries, the components could not be readily recycled into a successor machine. Of about 14 made, one is preserved in a display cabinet in Cambridge University and is regularly demonstrated in full working order to economics undergraduates (though not normally to the public). Another is on permanent display at the Science Museum in London, not in working order but with a video showing its operation. A third is in New Zealand, where it was incorporated into the country's display at the 2002 Biennale arts exhibition in Venice. A fourth lives on in the foyer of the University of Melbourne's Economics and Commerce department.

Australia recognized the importance of its first computer rather better than other countries, and when the CSIRAC was finally switched off in 1964 it was donated to the Institute of Applied Science of Victoria (now the Museum of Victoria) instead of being scrapped. The team in the computation laboratory at the University of Melbourne, where it had spent the last 10 years of its working life, knew that it was a historic machine, the longest-serving computer in the world and the only first-generation one still in use. So the deputy engineer, George Semkiw, packed it up very carefully, labeling everything, complete with its documentation, program tapes, and much else. It spent much of the next 30 years in storage, survived a flood that reached the base of the machine in 1995, and the following year was reassembled for exhibition in the Department of Computer Science at Melbourne. Peter Thorne and the other surviving veterans of the CSIRAC's operational years decided not to attempt to get it working again, as they would have had to renew a lot of components and it would no longer be original. Moreover, it is unlikely it could have been kept in working order "once we fell off our perch!" So it would have soon become unusable again: "We would have taken an original, non-working machine, and turned it into a non-working, non-original machine." It is now back in Victoria, again in storage but this time safe from flooding and viewable on request. After its two lives, first as a research machine at Sydney, then as a workhorse at Melbourne, it has now been reincarnated, in the words of Peter Thorne, as "an icon," symbolizing Australia's contribution to the creation of modern computing.

Among the British computers mentioned here, it seems the original EDSAC is no more and just a small fragment of the Manchester Baby is preserved, and that only because it was recycled into a support for a retaining wall and rescued decades later. However, a faithful replica of the latter using many contemporary components was built in time for the 50th anniversary celebration in 1998 and is now on display at the Museum of Science and Industry in Manchester. It is a working copy of the original and is demonstrated once a week. The Computer Conservation Society is doing a remarkable job in preserving and restoring a range of British computers from earlier decades and more details can be found on its website (*www.bcs.org/sg/ccs/*). Similar efforts continue in other countries.

The final legacy is, of course, the human one, the aspect that this book has concentrated on. LEO's lasting influence is typical, as Frank Land argued in 2001: "It was significant in acting almost as a university in disseminating how to do things. There are senior people all over the world who learned how to do it with LEO and themselves propagated this further down the line. So it had far more influence than the 100 or so machines we sold. This went all over the world. LEO people went to America, to South Africa, and they were prominent there, and they still are." In 1967 Land was recruited by the London School of Economics to set up a department in the new field of systems analysis. "I said farewell to LEO and became an academic—a move I have never regretted, nevertheless with strong feelings about what I had learned at LEO and with LEO. Certainly the basis of what I taught at the LSE was founded on what I learned during those very formative years. It was a marvelous time. It had its downs as well as its ups, but it was quite incredible what forward thinking took place here in Elms House and at Hartree House, all those years ago." Fifty years later, when new computer projects are still too often badly planned and imposed on an unwilling workforce, it is impressive to see how hard the LEO team tried to ensure that their computer installations made users' lives easier.

David Caminer, who played such a major part in the LEO project, stayed with the company as it metamorphosed into ICL, taking on ever more senior roles. He was awarded the OBE in 1980 in recognition of his work as project director of the multi-national team that implemented the communications and computer network for the European Community.

APPENDIX A

_____

BIBLIOGRAPHY AND FURTHER INFORMATION

This is not intended to be a complete list of all the books that have assisted me over the years, or even for this book, much of which is based on original interviews in any case. Rather all the books here come under the heading of "recommended reading" if you want to know more about the projects covered in each chapter. Similarly there is a massive store of information on the Web, much of it unfortunately of questionable authority and often regurgitating myths picked up elsewhere. However, some examples of useful sites are included. Just remember that even the most official and original-looking document will have been written for a reason and isn't necessarily the final word on the subject.

I can be contacted at *author@electronicbrains.com*.

PROLOGUE

The Cogwheel Brain: Charles Babbage and the Quest to Build the First Computer by Doron Swade (Little, Brown 2000). Published in the US as *The Difference Engine: Charles Babbage and the Quest to Build the First Computer* (Viking 2001).
"The Life and Work of Konrad Zuse" at *www.epemag.com/zuse* by his son Horst Zuse.

Pictures and descriptions of some original differential analyzers and a modern-day version built from Meccano can be seen at *www.meccano.us/differential_analyzers*.

1. FROM ABC TO ENIAC
The ENIAC material is based largely on original interviews recorded with

241

Jean "Betty" Bartik, Kay McNulty Mauchly Antonelli, Art Gehring, Max Kraus, Jim McGarvey, and Nathan Ensmenger in the autumn of 2001.

From ENIAC to UNIVAC by Nancy Stern (Digital 1981).
Atanasoff: Forgotten Father of the Computer by Clark Mollenhoff (Iowa State University 1988).
"The Origin of the Stored-Program Concept" by Allan Bromley, in *The Last of the First: CSIRAC Australia's First Computer* by Doug McCann and Peter Thorne (University of Melbourne 2000). An excellent analysis.
"The Unisys Newsletters" by George Gray at *www.cc.gatech.edu/services/unisys–folklore.*
The Pennsylvania University Library online exhibition on ENIAC: *www.library.upenn.edu/exhibits/rbm/mauchly/jwm11.html.*
The late Mike Muuss's "History of Computing Information" at *ftp.arl.mil/~mike/comphist/* is a valuable archive, an official collection of original US Army documents and reports about ENIAC, EDVAC, etc.
A remarkably well-constructed site at *home.att.net/~thercaselectron* shows how the Selectron worked.

2. UNIVAC—SAVIOR OF THE CENSUS

The UNIVAC material is largely based on the same interviews as Chapter 1. Most of the other Chapter 1 references are also relevant to this chapter. In addition the full text of Calvin Mooers's comments on Atanasoff can be found in "Calvin Mooers, the NOL Computer Project, and John Vincent Atanasoff: An Introduction" by Michael R. Williams of the University of Calgary, in the *IEEE Annals of the History of Computing*, April–June 2001 (this includes the Calvin Mooers memoir). Can also be found on the Web.

3. SALUTING THE MOOSE

Based largely on original interviews recorded in Connecticut, USA (2001), with local historian Erik Rambusch and Rand engineers Bill Wenning, John Carmichael, Gordon Chamberlain, Mike Norelli, Jim Marin, Cliff Beirle, and Les Henchcliffe. Greatly assisted by email correspondence (2004) with Andrew Egendorf, who has tracked down some rare documentation.

No books, or even chapters, have been published on the Rand 409 series.

The history of Rowayton referred to is *Rowayton of the Half Shell* by Frank E. Raymond (Phoenix Publishing 1990).
The Rowayton Historical Society can be found at *www.rowayton.org/rhs/Computers/welcome.html.*

4. WHEN BRITAIN LED THE COMPUTING WORLD

Based on an interview recorded with Sir Maurice Wilkes (Cambridge 2004), a range of other published sources, and wherever possible original documents, many of which can be found on the Web.

Alan Turing: The Enigma of Intelligence by Andrew Hodges (Hutchinson 1983).
   Much has been written about Turing, but none better than this biography.
Memoirs of a Computer Pioneer by Maurice Wilkes (MIT Press 1985).
Many original Turing documents can be found on the Web, with
   www.alanturing.net and *www.turingarchive.org* both excellent starting points.

5. LEO THE LYONS COMPUTER

Based largely on original interviews recorded in the UK (2001) with David Caminer, Peter Bird, Frank Land, Ralph Land, Mary Coombs (née Blood), Yvonne Dolezal, and Kathleen Bush.

Leo: The Incredible Story of the World's First Business Computer by David Caminer,
   John Aris, Peter Hermon, Frank Land, et al. (McGraw-Hill 1998).
   Originally published in the UK as *User-Driven Innovation: The World's First
   Business Computer.*
The First Food Empire: A History of J. Lyons and Co. by Peter Bird (Phillimore
   2000).
A Computer Called LEO by Georgina Ferry (Fourth Estate 2003).
LEO: The First Business Computer by Peter Bird (Hasler 1994).
The LEO Computers Society is at *www.leo-computers.org.uk.*

6. SO THEN WE TOOK THE ROOF OFF

The MESM and BESM accounts are based largely on original interviews recorded in Kyiv, Ukraine (2001), with Boris Malinovsky, Zinovy Rabinovich, Victor Ivanenko, Ivan Parkhomhenko, Rostislav Cherniak, and his daughter Svetlana Cherniak. Information on Bruk is based on an interview and a paper by Sergei Prokhorov (Moscow 2001). Overview provided in a recorded interview with Doron Swade (London 2001).

Computing in Russia by Georg Trogemann, Alexander Nitussov, and
   Wolfgand Ernst (Vieweg [Bertelsmann] 2001). A comprehensive and
   detailed history with contributions by a number of pioneers.
Red Computers: How Russia Lost the Computer Cold War by Boris Malinovsky
   (M. E. Sharpe 2005).
A comprehensive history of Ukrainian computing, in English, can be
   found at *www.icfcst.kiev.ua/museum/.*

7. WIZARDS OF OZ

Based partly on interviews and correspondence with Harry Messel, Peter Thorne, Kay Thorne, and John Deane (2004).

A History of Australian Computing by Trevor Pearcey (Chisholm Inst. 1988).
The Last of the First: CSIRAC, Australia's First Computer by Doug McCann and Peter Thorne (University of Melbourne 2000).
SILLIAC: Vacuum Tube Supercomputer by John Deane (Australian Computer Museum Society, not yet published).

8. WATER ON THE BRAIN

Based largely on original recorded interviews in the UK (2001) with Walter Newlyn, Nicholas Barr, Brian Henry, and Richard Lipsey, and an additional interview with Heather Sutton (2004).

A. W. H. Phillips: Collected Works in Contemporary Perspective edited by Robert Leeson (Cambridge University Press 2000). Includes his main academic papers but the first half is about Phillips, the man and the computer.

9. IT'S NOT ABOUT BEING FIRST: THE RISE AND RISE OF IBM

The Maverick and His Machine: Thomas Watson Sr. and the Making of IBM by Kevin Maney (Wiley 2003). Written with rare access to IBM archives.
Think: The Amazing Story of IBM by William Rodgers (Weidenfeld and Nicolson 1970). "Written with the active non-cooperation of the company," it says, but Thomas Watson Jr. has praised its accuracy though it pulls no punches.
Father Son & Co: My Life at IBM and Beyond by Thomas J. Watson Jr. and Peter Petre (Bantam 1990). A remarkably candid account.
Howard Aiken: Portrait of a Computer Pioneer by I. Bernard Cohen (MIT Press 1999).
Inside IBM: A European's Story by Jacques Maisonrouge (Collins 1988).
IBM and the Holocaust by Edwin Black (Little, Brown 2001).
The IBM website *www.ibm.com* has a wealth of historical information (and a response to Edwin Black's book).

EPILOGUE

The discussion on the relationship between World War II and the development of modern computing was prompted partly by the comments of many of the interviewees and also by the Open University course "War, Peace and Social Change."

For the general context (neither volume specifically considers the relationship between warfare and computing), see:

Total War and Social Change by Arthur Marwick (Macmillan 1988).
The People's War: Britain 1939–45 by Angus Calder (Pimlico 1992).

GENERAL

The Origins of Digital Computers: Selected Papers edited by Brian Randell (Springer 1982). A treasure trove of historical documents to do with analytical engines (e.g., Babbage), tabulating machines (e.g., Hollerith), Zuse and Schreyer, Aiken and IBM, early electronic computers (e.g., ABC) and stored-program computers (EDVAC, etc.).

The recorded interviews referenced above may become available in the future to bona fide researchers. For the latest information, see *www.electronicbrains.info.*

CONSERVATION

As recounted in the Epilogue, many of the early computers have been partly or totally lost. However, some artifacts and some complete computers are preserved and on display to the public.

There are also societies in a number of countries dedicated to preserving the history of early computers and where possible preserving or even restoring the machines themselves.

Andrew Egendorf in the USA is setting up a contact list of all those who worked on the early computers and are still around to talk about it, which means the first models, such as the original LEO, the first Rand 409s, the original UNIVAC, and so on. More details are available on the *www.electronicbrains.info* website.

LEO: Sadly little remains of any of the LEO computers, though some pieces are on display at the Science Museum in London. The history is well cared for by the vigorous LEO Computers Society (*www.leo-computers.org.uk*). The very first Lyons Teashop at 213 Piccadilly in London is now a café called Ponti's, where it is still possible to see the original Lyons ceiling. Elms House, where LEO II was developed, is now a record company office and not open to the public. However, it still looks the same from the outside and is the only recognizable remnant of Cadby Hall.

Rand 409: Amazingly all 1,500 Rand 409/UNIVAC 60/120s seem to have gone. A few artifacts, like some of the valve chassis units, are kept by the Rowayton Historical Society in Connecticut. The barn, where most of the development work was done, is still complete and in excellent condition, functioning as the community center, town hall, and library, and even the moose head remains.

ENIAC: Some of the 40 or so units were saved and are on permanent display at the University of Pennsylvania in Philadelphia. Von Neumann's IAS computer, which influenced so many others, remains at the Smithsonian Institution in Washington.

MESM: The first Soviet computer, too, didn't survive, but one of the great BESM series that followed was rescued by Doron Swade in the early 1990s and is in a Science Museum warehouse awaiting restoration. Feofania at the edge of Kyiv remains a lovely area, the Secret Laboratory No. 1 is now a convent, and the adjacent church has been meticulously restored and is again in daily use for worship. There is an excellent display of the history of calculators and computing at the Polytechnical Museum in Moscow (the city's oldest museum), including part of a BESM.

EDSAC and others: While EDSAC has gone, a complete NPL "Pilot ACE" is on display at the Science Museum, and the Computer Conservation Society (*www.bcs.org/sg/ccs*) is working to restore a number of early computers. A working replica of the Manchester Baby was built for the 50th anniversary of its first run and has found a home at the city's Museum of Science and Industry. The code-breaking computers have of course received a great deal of attention at Bletchley Park. Astonishingly a working Colossus has been rebuilt, largely from memories and photographs, while a Bombe is also being built with the intention that it too will work (visitors can view both).

CSIRAC and SILLIAC: The Australians took a bit more care of their first computer than other countries and the complete CSIRAC still exists despite some precarious years (see Epilogue). CSIRAC is viewable on request at the Museum of Victoria.

Phillips (Moniac): Considering how few of these were made, and their intrinsic fragility, the Phillips machines have survived rather well. Only one, at Cambridge, is known to be still working. It is not on public display, but is still demonstrated to economics students. A display model can be found at the Science Museum; it has been "embalmed" to preserve it, so a video of it in operation plays on an adjacent console. Another is on display in the first-floor lift well of the Economics and Commerce building at Melbourne University in Australia, while yet another has been proudly hosted at New Zealand's Institute of Economic Research in Wellington since 1987.

ARITHMETIC

DECIMAL AND BINARY ARITHMETIC

The fundamental breakthrough in moving from passive mathematical aids (such as the abacus) to calculating machines was the realization that much of mathematics could be expressed in mechanistic steps. No computer yet invented can be given Fermat's Last Theorem and produce a proof for it. But the calculation of log tables, star positions, and tides, and the solution of simultaneous and differential equations, can all be broken down into simple arithmetical steps. So a machine that is capable of addition and subtraction can, if necessary by repetition of the same step thousands of times over, also perform multiplication and division, and hence even more complicated mathematics.

Addition is straightforward, provided the machine can add two digits together and "carry" any overflow, just as a child does basic sums. The same applies to subtraction.

Multiplication is a little different. A human would approach a calculation such as $1,732 \times 48,977$ by setting it out as follows, using "times tables":

$$
\begin{array}{r}
48,977 \\
\times\, 1,732 \\
\hline
97,954 \\
1,469,31\_ \\
34,283,9\_\_ \\
48,977, \\
\hline
84,828,164
\end{array}
$$

The computer doesn't need to know the times tables. It just adds 48,977 to 48,977 over and over again, a total of 1,732 times, to reach the same answer. That would be unimaginatively tedious, time-consuming, and error-prone for a human to do, but even the early electronic computers could do it in a fraction of a second.

Division is similar. You want to divide 48,977 by 1,732? Just subtract 1,732 from 48,977 and repeat until the remainder is less than 1,732, counting the number of times. The result will be 28, with a remainder of 481. To get the first decimal place, multiply the remainder by 10 and repeat the process and so on until you have the accuracy you need.

Some early computers, such as the ENIAC, did calculate in decimal and by running the machine slowly you could actually see from the glowing valves the very steps outlined above. However, most computer designers quickly adopted the base-2 or binary system, which is inconvenient for mental arithmetic but more efficient for machines.

Take, for example, the addition of two numbers, in decimal:

$$\frac{\begin{array}{r} 433 \\ 213 \end{array}}{646}$$

The same numbers in binary are added in the same way:

$$\frac{\begin{array}{r} 0110110001 \\ 0011010101 \end{array}}{1010000110}$$

To most of us utterly familiar with decimal numbers (even if multiplication is a challenge) the binary version looks daunting and rather meaningless. We have some idea how big 433 is, but 0110110001 could be anything and binary requires a lot more digits to express a given quantity. However, a computer sees things differently. Each decimal unit needs 10 switches, to represent the numerals 0, 1, 2, 3 . . . 9. So the decimal addition needs 30 switches for each 3-digit number. However, a binary digit needs only 1 switch ("off" means "0" and "on" means "1"), so each of the 10-digit numbers above requires only 10 switches.

That's a good enough reason in itself to opt for binary (even allowing for the need to convert decimal numbers to binary and back again to keep the poor human operators happy). But when you look at the problem of multiplication it gets even better:

$$433$$
$$\times 213$$
$$\overline{????}$$

Multiplication is difficult for a decimal computer. One obvious way is to think of it as repeated addition. So you add 433 to itself and repeat another 211 times. And that is how early computers like the ENIAC worked.

But in binary arithmetic, multiplication works by the following simple rule:

$$0 \times 0 = 0$$
$$0 \times 1 = 0$$
$$1 \times 0 = 0$$
$$1 \times 1 = 1$$

Conveniently that is also the result from a simple electronic circuit called an AND-gate. So multiplication of say 9×3 becomes:

$$1001 \quad \text{[decimal 9]}$$
$$\times 011 \quad \text{[decimal 3]}$$
$$\overline{1001}$$
$$1001$$
$$\overline{11011}$$

If you convert all that back to more familiar decimal digits you'll see that $9 \times 3 = 27$. And that calculation can be done neatly in the computer by a combination of AND-gates and simple addition.

Is "binary" the same as "digital"? No, although the two often go together. An analog quantity is one that varies continuously, as in the example cited earlier of a traditional glass thermometer, where the mercury rises and falls smoothly to indicate temperature. A modern electronic thermometer shows the temperature as numbers—digitally. This can give a very accurate display but it still changes in steps, as the digits change.

Another way of looking at the contrast between analog and digital is to consider a modern music system. The sound of the singer starts off as an analog signal—varying continuously—is picked up by a microphone (still analog) and at some point in the recording system goes through an analog-to-digital converter on its way to eventually being recorded on a (digital) compact disc. When you play the CD in your sound system the process is reversed. The digital signal from the CD goes through a digital-to-analog converter and ultimately (with some amplification along the way) to your loudspeakers, which are analog devices.

The binary system of arithmetic is ideal for a digital computer. The computer counts in steps, because it is digital, and as was shown above those steps can be most efficiently handled by an electronic computer.

SIMULTANEOUS EQUATIONS

The drive for computers was driven by the need to solve complex mathematics. One complex problem was simultaneous equations. Many activities in the physical world can be modeled by large arrays of simultaneous equations but they take a long time to solve. However, if they are simple linear equations, it is an easy job even for an early computer. It requires a very large number of computational steps, but at electronic speed that is not a great problem—it is still much quicker (and more accurate) than a human being could manage. Even for the non-mathematically minded it's not complicated to understand.

Take the simplest equation with one variable: $3a = 9$. To put it in everyday terms, that might be saying that the cost of three apples is nine cents. So simple arithmetic says one apple is three cents.

Now take two equations:

$$a + 11b = 14$$
$$4a + 7b = 19$$

So what are a and b? Well, you could multiply both sides of the first equation by 4 and it would still be true (because you multiplied both sides by the same amount) and it would give you:

$$4a + 44b = 56$$
$$4a + 7b = 19$$

Now subtract the second equation from the (modified) first equation and you get:

$$4a + 44b = 56$$
$$\underline{4a + 7b = 19}$$
$$0 + 37b = 37$$

so clearly $b = 1$.

But it was quite intelligent to notice that you could simply multiply one equation by 4 and subtract the other from it. Computers then, even now, don't have that kind of intelligence. Again there's a more mechanical approach that suits a computer. Take our two equations again:

$$a + 11b = 14$$
$$4a + 7b = 19$$

Now from the first equation, what does a equal? It equals $14 - 11b$. Not too useful as you don't know what b equals. But substitute $a = 14 - 11b$ in the second equation and that gives:

$$4 \times (14 - 11b) + 7b = 19$$

Multiply the numbers in the brackets and

$$56 - 44b + 7b = 19$$

Do the addition and subtraction and

$$56 - 37b = 19$$
$$56 - 19 = 37b$$
$$37 = 37b$$

So $b = 1$.

Now the great thing about this method, known simply as "substitution," is that it keeps working. If you have say six variables, a, b, c, d, e, f, then provided you also have at least six unique simultaneous equations, they can be solved by the substitution method. It quickly gets tedious, but computers don't get bored (unless they're in *The Hitchhiker's Guide to the Galaxy*).

This was what the Atanasoff Berry Computer was designed to do: in fact it had sufficient memory to solve an array of 29 simultaneous equations each with 29 variables. A daunting prospect for a human mathematician, and the sort of thing that took up weeks of Maurice Wilkes's Ph.D. work before he built the EDSAC.

DIFFERENTIAL EQUATIONS

This was another fundamental problem, and the best answer until the ENIAC came along was the Bush differential analyzer. But the mechanical device was analog and its accuracy depended on constant adjustments by the operator and mechanics. Again a digital computer (binary or decimal) could solve some classes of differential equation.

Unfortunately solving differential equations is rather more complicated than the simultaneous equations used in the examples above. So it is not explained here.

BI-QUINARY ARITHMETIC

This is a variation of decimal arithmetic and the basic building block of what the Rand 409 team called "Crosman's logic." This was a decade counter, like the ENIAC (and the Colossus) rather than the straight binary circuits used in most other early electronic computers. A conventional decimal counter would simply count from 0 to 9 and then "carry" 1 to the next counter to indicate "10," and so on. "Bi-quinary" meant that each counter would increment from 0 to 4, then "carry" a 5 and count up from 0 to 4 again (representing the numbers 5 to 9), before carrying 1 to the next counter (to represent 10). It's a rather complicated mathematical process, but it was quite efficient and in fact the traditional abacus used the same system.

APPENDIX C

TECHNICAL BITS

TYPES OF "COMPUTER"

One of the fundamental problems with the question "Who invented the computer?" is that it immediately invites the question "What do you mean by computer?"

The first computers were people, usually women and usually spelled "computor." They were trained mathematicians who carried out complicated calculations, making use of whatever mechanical aids were available at the time, such as slide rules and differential analyzers. A good example is the roomful of 100 or so "computors" who slaved over gunnery firing tables during World War II and led to the ENIAC project as recounted in Chapter 1.

It wasn't until the 1950s that "computer" became the generic term we know today. In the late 1930s and throughout the 1940s, machines that might qualify for that label were called "calculating machines," "computing machines," "automatic calculators," and so on. Even as late as 1951, Maurice Wilkes's published papers referred to "electronic calculating machines."

Understandably writers in later decades, when decoding the acronyms that were given to most computers, have frequently assumed that the "C" (in EDSAC, for example) always meant "computer," whereas it often meant "calculator." However, while it is useful to know the correct title for any of these historic machines, that doesn't help decide which was truly a "computer." Indeed there is no universally accepted definition.

Babbage's "engines" were computers of a kind but wholly mechanical,

253

though they included some remarkably modern concepts, such as programmed operation by punched cards, and conditional branching (the ability to use the result of one calculation to decide which calculation to perform next). Turing's 1936 "universal machine" was also a computer of a sort but a purely hypothetical construct intended to prove a mathematical conjecture.

Konrad Zuse's Z1 to Z4 used electromechanical relays for calculation at mechanical rather than electronic speed, so they too are not generally regarded as true modern computers. Howard Aiken's Harvard machines, too, used primarily mechanical calculation.

With the ABC we come close to a true modern computer. It was electronic, even if it wasted a lot of time (relatively) waiting for its mechanical memory drum (full of capacitors) to trundle around to the right location for reading and writing data. It used binary numbers and it calculated by electronic logic rather than simple "enumeration." It pioneered the storage of data in capacitors, and the use of regenerative pulses to stop the charges (and hence the data) "leaking" away, and that is still the basis of "dynamic RAM." But it never fully worked and its development came to an end when Atanasoff went on to war work. If it hadn't been for the connection with Mauchly it would have had no influence at all on subsequent computer development and in all probability would have been forgotten forever (instead of just until the 1970s). Nonetheless it can be claimed as "the first digital electronic computer," though it was neither a general-purpose nor a stored-program computer.

The mid-1930s Polish "Bomba" and its British "Bombe" successor were more mechanical computers, of a very specialist kind. Much more interesting was the Colossus (1943). This was electronic, and in its Mark II version had conditional branching. It had a working memory though the program wasn't stored in it. It was specifically designed for code breaking and although it could be adapted to other computational work it wasn't a truly general-purpose computer. So, although it pre-dated the ENIAC by two or three years, it probably loses out to the latter, which can reasonably be claimed as "the first general-purpose digital electronic computer." Note, however, that the ENIAC was decimal not binary, and never had a true stored-program capacity.

Only after World War II do we see the emergence of the first true modern computers, using what is now so well known as "von Neumann architecture," the most significant distinguishing feature being memory that contains both data and program, which can thus (at least in principle) be modified by the computer itself. The EDVAC, EDSAC, UNIVAC,

CSIRAC, LEO, ACE, MESM, etc., were all stored-program computers, and indeed that, as we've seen, was one of the major development challenges. In 1948 the Manchester Baby was the first to successfully run a program from memory shared with data, but—and there's always a "but"—it was a prototype with minimal processing ability designed purely to test whether that kind of memory (a cathode-ray tube) would work in a computer. Within a year the EDSAC too ran for the first time, though that team had to be content with the claim of "first digital electronic stored-program computer built for general use."

The requirement for stored-program capacity trips up the Rand 409, which was programmed by plug boards. Yet, as a general-purpose punched-card computer, it was capable of a host of business applications such as tax calculation and payroll, and of course it sold in large numbers. Whether it can claim to be the world's first business computer depends as always on definition. Certainly Loring Crosman conceived it remarkably early, apparently in about 1943, and the first delivery was in mid-1951. But that was to the Internal Revenue, and some would argue this was not a true business application. The IBM Card-Programmed Calculator went on the market as a business machine in 1949, but this was a curious hybrid of the company's 604 Electronic Calculator, 402 Electronic Accounting Machine, and an external relay memory. Because it executed instructions directly from external punched cards it didn't operate continuously at full electronic speed and by some definitions that debars it from being considered a true electronic computer. The LEO computer, the other main contender for the title of "first business computer," ran working business programs in 1951, some time before the Rand 409 got into payroll departments. Moreover, LEO, being based on the EDSAC, was a true stored-program computer, and the complexity and elegance of its wide range of applications reflected that extra capability over the Rand model.

With the exception of the Rand 409, virtually all electronic computers from the Baby onwards used von Neumann architecture (or something very similar), including the stored-program concept.

RANDOM-ACCESS MEMORY, OR RAM

Almost anyone who uses a personal computer today knows the word RAM, but not everyone understands what it is. Early computer memories, like the mercury tank acoustic delay-lines described in earlier chapters (and used in machines like the EDSAC and LEO), were not "random-access memories." A long series of data bits circulated around the memory system and the computer frequently had to wait for the right piece of data to appear before it could process the next instruction. Several other early memories, like the rotating magnetic drum, were also slow to use because they could not be accessed at random. By contrast, data stored in the Williams–Kilburn Tube (used in the Manchester Baby) could be accessed at random, because the beam that "read" the data could be switched to any part of the screen where the data were stored. Thus there was no waiting for the right data to appear and the computer could operate faster. Hence the "random access" became an important feature of such memory and the name lives on, though almost invariably in the abbreviated form "RAM."

THE GAME OF "NIM"

This ancient game (also known as "fiddlesticks') was based on a simple idea that was easy to simulate on early computers and thus proved popular at Melbourne University's open days when they showed off the CSIRAC to the public.

Two players have two or more piles of counters of some sort (stones, beads, sweets, etc.). They take turns to remove any number of counters from one pile at a time. The winner is the person to take the last counter. It's not as easy as it seems, though there are winning strategies. There are many simulations on the Web, though often they have been programmed to make sure the computer always wins!

INDEX

Numbers in *italics* indicate captions.